Geology of Titanium-Mineral Deposits

Eric R. Force
U.S. Geological Survey, Tucson Field Office
Gould-Simpson Building
University of Arizona
Tucson, Arizona 85721

SPECIAL PAPER
259

1991

Published by The Geological Society of America, Inc.
3300 Penrose Place, P.O. Box 9140, Boulder, Colorado 80301

GSA Books Science Editor Richard A. Hoppin

Printed in U.S.A.

Library of Congress Cataloging-in-Publication Data
Force, Eric R.
    Geology of titanium-mineral deposits / Eric R. Force.
        p.   cm. — (Special paper / Geological Society of America ;
    259)
    Includes index.
    ISBN 0-8137-2259-4
    1. Titanium ores.   2. Titanium dioxide.   I. Title.   II. Series:
Special papers (Geological Society of America) ; 259.
TN490.T6F67   1991
553.4′623—dc20   90-23807
                                                                CIP

Front cover: Blue Seascape Wave Effect, painted by Georges Lacombe (1868–1916), held by Musee des Beaux-Arts, Rennes/ Giraudon, Paris/Bridgeman Art Library, London. The painting accentuates two stages in a breaking wave. To the left, a freshly broken segment delivers sediment suspended in turbulent water, and deposits a suspension-equivalent layer as it decelerates up the swash face. To the right are a yet-unbroken segment and the nearly stationary remnants of the previous wave, its former turbulent cells delineated by domains of bubbles. The return swash is a sheet flow with a basal viscous boundary layer, and entrains coarse particles. The successive action of the two flow types results in placer enrichment in shoreline deposits of titanium minerals. See Figure 62 and pages 62–63, 74–75.

10   9   8   7   6   5   4   3   2

Contents

Contents

Igneous Processes

Weathering

Sedimentation

Preface

The mining of titanium minerals is one of the few sectors of the mining industry that suffered no downturn in the 1980s. Indeed, prices of titanium minerals are climbing, exploration is proceeding, and new mines are opening. Substitutes for titanium minerals in either pigment or metal applications seem remote. Some observers claim that titanium is one of only a few metals whose use will increase in the near future.

However, students of the supply picture warn that placer resources of titanium minerals are being depleted and that their replacement by other placer deposits is unlikely. Thus the economic geology of titanium minerals is in flux, and few questions about the future of titanium-mineral supply can be answered with confidence. By the year 2010, we are likely to see titanium minerals being produced from deposit types not currently exploited.

Thus, there is a need for a treatise on the economic geology of titanium that covers all the basic processes of formation of titanium-mineral deposits. No such treatise has ever been published. The opportunity attracted me because after 18 years as titanium resource specialist for the U.S. Geological Survey I am sufficiently familiar with the many types of deposits that I can address the whole topic. Indeed, I see some genetic relations between deposit types that can adequately be treated only in this comprehensive way. One of my hopes is to provide an integrated reference work usable by all geologists interested in titanium minerals. Clearly, parts of it will be of interest primarily to economic geologists. My emphasis, however, is on geologic processes of titanium oxide-mineral formation and concentration. For this reason, the volume may be of interest to other geologists oriented toward earth processes, such as geochemists and sedimentologists. At the center of the audience addressed by this approach is the future explorationist. Personally, I have found viewing the realm of geologic process through a "titanium window" to be rewarding in its own right, independent of economic considerations. The process emphasis is also the reason I chose the Geological Society of America as publisher.

This book is organized along the lines of a geochemical cycle of titanium, in order to facilitate the description of linkages among deposit types. Metamorphic and igneous processes of titanium oxide-mineral formation and concentration are described first (Chapters 1 through 5), followed by a description of their passage through the weathering process (Chapter 6) and their behavior as detrital particles in sedimentation (Chapters 7 through 10).

I would like to recognize and thank my field guides and intellectual partners—Are Korneliussen, Mike Shepherd, Langtry Lynd, Andrew Grosz, Curt Larsen, Ragnar Hagen, Norman Herz, Sherman Marsh, Gerry Czamanske and Bill Moore, Michael Foose, Bruce Carter, Fred Rich, Tom Garnar, Pierre Marcoux, Suzie Nantel and Michel Hocq, Paul Ollila, Allan Kolker, Dick Goldsmith, Jim Minard, Theo van Leeuwen, Raphael Alexandri, Jeanette

Hilbish, Warren Anderson, P. A. Lindberg, and Byron Stone. Figure 2 is the idea of Gilpin Robinson. Reviewers other than those listed above include John Hosterman, Steve Haggerty, Frank Lesure, Peter Roy, Malcolm Ross, and Dave Gottfried; Cheryl Varner edited the manuscript. I was accompanied in the field at many described deposits by my wife, Jane. The manuscript of this volume and its many revisions were typed by Iris Howard. This volume was written, and my research for it done, while I was a member of the Branch of Eastern Mineral Resources at the Reston, Virginia, headquarters of the U.S. Geological Survey.

Geological Society of America
Special Paper 259
1991

ABSTRACT

More than 90 percent of the titanium minerals currently produced come from magmatic ilmenite deposits and from young shoreline placer deposits. This means that the two geologic processes most directly responsible for economic titanium-mineral deposits are (1) the accumulation of dense oxide-rich liquids immiscible in cooling magmas of ferrodioritic to gabbroic composition, and (2) the interference between deposition and entrainment in the enrichment of dense minerals on the upper swash zones of beaches (and removal of some concentrates to eolian environments). Both processes are essentially mechanical; i.e., chemical remobilization of titanium does not form its major ore deposits.

Both processes also require precursor conditions that ensure that titanium is present predominantly in the form of oxide minerals. In magmatic deposits, these are physical and chemical conditions that favor titanium-oxide over titanium-silicate minerals. In sedimentary deposits, these conditions are a combination of proper source rocks, weathering history, and sedimentary conduits, all necessary to permit the supply of favorable minerals and prevent their dilution with unfavorable ones.

Some titanium-mineral production currently comes from fluvial placer deposits (Gbangbama, Sierra Leone) and from deeply weathered alkalic pyroxenites (Tapira, Brazil). In addition, several other deposit types could well become economic in the near future: (1) rutile from eclogites, (2) rutile from contact-metasomatic zones of alkalic anorthosites, (3) perovskite from alkalic pyroxenites, and (4) rutile byproduct from porphyry Cu-Mo deposits; detrital titanium-mineral deposits could be exploited (5) on continental shelves, (6) in Pleistocene glaciolacustrine deltas, or (7) in older, semiindurated beach deposits. If young shoreline placers are depleted, these other deposit types may become important.

Chapter 1.

Introduction
The impact of titanium-mineral geology on industry and trade

Titanium minerals are currently mined from four quite dissimilar types of deposits. To a great extent, the structure of the titanium industry reflects the mineralogy of the different deposit types and subtypes, and world trade patterns for titanium products reflect deposit-type distribution. This chapter explores the relation between geology and industry; for this purpose the economic geology of titanium is introduced in nongenetic terms.

USES AND VALUE OF TITANIUM MINERALS

Titanium metal and TiO_2 pigment are the two main products made from titanium minerals, and on them large industries depend. First by far in terms of volume produced is microcrystalline TiO_2 for white pigment. Because of the extremely high refractive index of TiO_2 as rutile (2.6 to 2.9, or higher than diamond), it is the chief opacifying pigment used in paint and other products such as plastics and paper, not only for white color but for quite a range of colors. It has supplanted lead-based pigments in many of these roles. Titanium dioxide pigment commonly forms more than 20 percent by weight of some paints. The pigment industry consumes more than 90 percent of all titanium minerals mined.

The product ranked second by volume, though perhaps not in importance, is titanium metal. The high strength-to-weight ratios and resistance to corrosion and high temperature of titanium metal and titanium-based alloys make them important ingredients in many industries. Most important is the aircraft industry, where the use of titanium has been growing for more than 30 years, to the point that commercial airliners of the current generation can contain 30 percent titanium by weight. Another trend during the same period has been a diversification of titanium-metal use in other industries. Many industries take advantage of titanium's corrosion resistance, as in heat exchangers and desalinization plants. The U.S. National Materials Advisory Board (1983) has published an extensive review of titanium metal use. Titanium is one of the few metals the use of which is likely to increase in the near future (U.S. Bureau of Mines, 1986).

The titanium industry is an infant compared with long-established industries using many other minerals and metals. Large-volume pigment uses date from the late 1930s, the metal uses from the late 1950s. The history of large commercial markets for titanium metal has been short but irregular, making normal market cycles look placid by comparison; fitful U.S. government acquisitions of military aircraft have partially governed the industry.

Worldwide, titanium-mineral mining in 1987 produced about 5.8 million metric tons of titanium-mineral concentrates (Lynd, 1988), worth approximately US $915 million. Titanium dioxide pigment and metallic titanium, made from titanium minerals, in the same year were worth about US $4,200 million and US $700 million, respectively. Total employment in these titanium-mineral mining and conversion industries is more than 20,000 workers. Past this point, it is impossible to track the increasing value of titanium products as they become finished goods in complex industries.

DISTRIBUTION OF TITANIUM AND TITANIUM MINERALS IN THE EARTH'S CRUST

Titanium is sufficiently abundant in the earth's crust (0.86 percent Ti, or 1.4 percent TiO_2, according to Turekian, 1977) that it is customarily listed with the major elements. Table 1 lists the TiO_2 contents of some important rocks; note that many common rocks have TiO_2 contents of more than 1 percent. Titanium is thus fundamentally unlike many mineral commodities that are present in only trace quantities in common rocks.

Titanium is present in rocks as oxide and silicate minerals. The minerals that account for most of the titanium in rocks are

TABLE 1. TYPICAL TITANIUM CONTENTS AND TITANIUM PARTITIONING OF SOME COMMON ROCKS*

Rock types	TiO_2 (%)	TiO_2 in oxide minerals (as % of total TiO_2)
Igneous rocks		
Ultramafic	0.2–0.8	1–4
Mafic	0.9–2.7	50
Felsic	0.2–0.8	3–30
Alkalic	0.1–3.3	3–50
Charnockitic	0.4–1.6	50–95
Anorthositic	0.1–0.5	50–95
Metamorphic rocks[†]		
Gneiss	0.6	5–100
Schist and phyllite	0.6	1–70
Amphibolite	1.4	15–70
Serpentinite	0.0	n.a.
Eclogite	1.0–6.0	50–90
Sedimentary rocks		
Sandstone	0.2–0.6	10–100
Shale	0.6–0.7	?
Limestone	0.1–0.2	?

*Modified from Force, 1976a.
[†]See Chapter 2 for variation of partitioning with grade.

listed in Table 2. The partitioning of titanium between oxide and silicate phases varies greatly among different rock types (Table 1). The economic geology of titanium begins by focusing on rock suites that contain titanium largely in oxide-mineral form.

INTERPLAY OF GEOLOGIC AND ECONOMIC FACTORS

Only the titanium-bearing oxide minerals have economic value. All the oxide minerals containing more than about 25 percent TiO_2 have some present or potential economic value, and no silicate minerals are valuable regardless of TiO_2 content. Thus the economic geology of titanium is the geology of titanium oxide minerals.

In addition, different titanium oxide minerals are best suited for different industrial recovery processes. The market price of a titanium-oxide mineral concentrate is a function of its TiO_2 content and its suitability for a given process. Prices of such concentrates vary an entire order of magnitude.

Therefore, mineralogy is a more important factor in the economic geology of titanium than it is for most other mineral commodities for which chemical enrichment is the most important factor. Examples will be given in this volume (Chapter 9) of important titanium-mineral ores that contain less than the average crustal abundance of titanium.

Economic minerals

The currently economic titanium oxide minerals are rutile, anatase, and ilmenite. Rutile, with a theoretical composition of pure TiO_2 (Table 2), is the most valuable, currently at about US $600 or more per metric ton. Two polymorphs of rutile—anatase and brookite—have the same theoretical composition, but they commonly contain detrimental chemical impurities. Anatase concentrates are just beginning to come onto the market, and if of suitable composition, they may also command a high price. A value for perovskite has not yet been established; it will depend on the efficiency of the process eventually used to convert it to TiO_2. Titaniferous magnetite in the strict sense has no value in the titanium industry at present. Silicate minerals having high titanium contents, such as sphene, have no economic value either.

Ilmenite, the most important source of titanium products, has a complicated story both geologically and economically. Unweathered ilmenite is commonly intergrown with iron oxide minerals and thus contains less TiO_2 than its theoretical composition indicates (Table 2). On the other hand, weathering leaches iron from ilmenite, resulting in poorly crystalline mineral grains residually enriched in TiO_2. The term *ilmenite,* as used in the titanium-mineral industry, commonly covers the entire range from unweathered ilmenite with TiO_2 contents below 50 percent to altered ilmenite containing more than 60 percent TiO_2. When the TiO_2 content of altered ilmenite exceeds about 70 percent, it is commonly referred to as *leucoxene.* Price per metric ton varies over this range from as much as US $500 for "leucoxene,"

TABLE 2. COMPOSITION OF SOME COMMON TITANIUM MINERALS

Mineral	Theoretical formula	TiO_2 content (%)
Oxides		
Rutile	TiO_2	>95
Anatase	TiO_2	>95
Brookite*	TiO_2	>95
Ilmenite	$FeTiO_3$	52[†]
Perovskite*	$CaTiO_3$	59[†]
Magnetite	Fe_3O_4	0–15
Silicates		
Sphene	$CaTiSiO_5$	41[†]
Melanitic garnet*	$Ca_3Fe_2Si_3O_{12}$	0–17
Biotite	$K_2(Mg,Fe)_4(Fe,Al)_2$ $Si_6Al_2O_{20}(OH, F)_4$	0–6
Calcic amphiboles	$(Na,K)Ca_2(Mg,Fe,Al)_5$ $Si_6Al_2O_{22}(OH,F)_2$	0–10
Augite	$Ca(Mg,Fe)(Si,Al)_2O_6$	0–9

*Phases restricted to unusual rock types.
[†]Stoichiometric value.

TABLE 3. TYPES OF TITANIUM-MINERAL DEPOSITS, THEIR ECONOMIC SIGNIFICANCE, AND THEIR TITANIUM MINERALS

Class	Type	Typical Mineralogy	Importance*	Example
1. Metamorphic	a. Eclogite	Rutile	B	Piampaludo, Italy
	b. Aluminosilicate	Rutile	E	Evergreen, Colorado
	c. Ultramafic contact	Rutile	E	Dinning, Maryland
2. Igneous	a. Magmatic ilmenite	Ilmenite	A	Allard Lake, Canada; Roseland, Virginia (in part)
	b. Anorthosite-margin	Rutile, ilmenite	C	Roseland, Virginia (in part)
	c. Albitite (kragerite)	Rutile	E	Kragerø, Norway
	d. Alkalic	Perovskite, Nb-rutile, Nb-brookite	C	Powderhorn, Colorado; Magnet Cove, Arkansas
3. Hydrothermal		Porphyry Rutile	C	Bingham, Utah
4. Sedimentary	a. Fluvial	Ilmenite, rutile	A	Gbangbama, Sierra Leone
	b. Glaciolacustrine	Ilmenite	C	Port Leyden, New York
	c. Shoreline (and coastal eolian)	Ilmenite, altered ilmenite, rutile	A	Richards Bay, South Africa; Stradbroke Island, Australia; Trail Ridge, Florida (in part)
5. Weathered	a. Alkalic parent rock	Anatase	B	Tapira, Brazil
	b. Mafic parent rock	Ilmenite	D	Roseland, Virginia (in part)
	c. Placer parent	Altered ilmenite, "leucoxene"	A	Trail Ridge, Florida (in part)

*A, of great present importance; B, of probable great importance in near future; C, of possible importance; D, of moderate present importance; E, of minor present importance at world scale.

through about US $70 for slightly altered ilmenite containing about 54 percent TiO_2, to ilmenite with lesser TiO_2 contents that is generally not traded (i.e., it is mined only by companies that consume it in their own pigment plants and smelters). Thus, the alteration state of ilmenite is of great economic importance. In the ilmenite mining industry, the term *grade* commonly refers to the TiO_2 content of ilmenite concentrates rather than to the amount of ilmenite in ore.

Deposit types

Titanium minerals are mined from hard crystalline rocks, weathered rocks, and unconsolidated sediments. Table 3 shows the great diversity of deposit types and the relations among them. These deposit types are described in detail in following chapters.

At present, shoreline placer deposits supply more than half the titanium minerals mined worldwide. These deposits supply rutile and variably altered ilmenite. Most of the remainder is supplied by magmatic ilmenite deposits from rocks of the anorthosite-ferrodiorite suite, which supply unaltered ilmenite. A fluvial placer deposit produces rutile, and a deposit in the weathered mantle on alkalic igneous rock has started producing anatase.

Currently known shoreline placer deposits will be exhausted in about 30 years, with several exceptions (Garnar, 1978; Shepherd, 1986; Fantel and others, 1986). The future importance of shoreline placer deposits is in question, and thus the economic geology of titanium minerals is in flux. New types of deposits would supply their own characteristic suite of titanium minerals (Tables 3, 4).

Recovery processes and relation to geology

The chloride and sulfate processes are the two recovery processes used in the titanium mineral industry. The chloride process converts titanium minerals to titanium tetrachloride and thence to either TiO_2 pigment or to titanium metal. This process is the more recent of the two and is preferred because it is less polluting. However, it requires a high-TiO_2 feed (and certain trace-element maxima). Initially, only rutile and leucoxene (TiO_2 >70 percent) could be used in the chloride process, but some companies have learned to use feeds of altered ilmenite containing about 60 percent TiO_2. Accordingly, plants using the chloride process are supplied predominantly with concentrates from weathered shoreline placer deposits. Chloride plants are also fed by placer rutile deposits of fluvial origin and could possibly use

**TABLE 4. WORLD PRODUCTION, RESERVES, AND IDENTIFIED
RESOURCES OF TITANIUM MINERALS**
(in thousand metric tons of contained TiO_2*)

Country	1987 Production[†]	Reserves[§]	Identified Resources[**]
Australia	1,250	27,000	131,000
Brazil	30	54,200	150,000
Canada	890	24,000	81,000
China	80	28,500[‡]	38,000[‡]
Finland	....	1,400	4,000
India	100	32,700	79,000
Italy	....	?	9,000
Malaysia	240	?	?
Mexico	....	....	3,000
Mozambique	....	....	2,000
New Zealand	....	....	47,000
Norway	550	29,000	89,000
Sierra Leone	110	1,800	2,000
South Africa	680	37,700	58,000
Sri Lanka	80	4,300	5,000
U.S.A.	360	10,600	103,000
U.S.S.R.	210	8,000[‡]	16,000[‡]
Total	**4,580**	**259,200**	**817,000**

*....... = none reported; ? = unknown.
[†]Lynd, 1988.
[§]Modified from Lynd, 1985.
[**]Including reserves. Fantel and others, 1986; Table 5 of this paper; and my information.
[‡]Reserve and resource figures are difficult to calculate for these countries because the Force and Lynd (1984) resource definitions do not apply. Figures listed by Towner and others (1988) are much larger.

anatase from weathered alkalic deposits or rutile from metamorphic deposits.

The sulfate process digests titanium minerals in sulfuric acid enroute to recrystallization as TiO_2 pigment. Effluents from this process are powerful pollutants unless they are neutralized. High-TiO_2 feed is not necessary in this process; indeed, the higher-TiO_2 feeds such as rutile and leucoxene are unreactive in it. Accordingly, plants using the sulfate process are fed by ilmenites containing 45 to 60 percent TiO_2. The trace elements acceptable in the sulfate process are also different from those acceptable in the chloride process. Most of the ilmenite used in the sulfate process comes from deposits in crystalline rocks and is unaltered. Some also comes from shoreline placers that are little weathered.

Two subprocesses are used to convert a low-TiO_2 feed into a high-TiO_2 feed for the two main recovery processes. First, in the smelting subprocess, low-TiO_2 feeds of appropriate composition (i.e., low Ca content but as little as 30 percent TiO_2) are smelted to a high-TiO_2 slag plus pig iron. This slag has a much higher market price than ilmenite (about US $275 per metric ton for 80 percent TiO_2 slag from Canada and US $300 for 85 percent TiO_2 slag from South Africa). Some of these slags are

suitable for the chloride process or, if processed via sulfate, greatly reduce the volume of effluent. In the second subprocess, "synthetic rutile" is produced from ilmenite by a number of methods; generally, feeds containing about 55 percent TiO_2 from slightly weathered placer deposits are used. The price of synthetic rutile, like that of rutile, is about US $600 per metric ton. Synthetic rutile is used in the chloride process. In fact, one producer is making titanium metal from ilmenite via synthetic rutile. In the near future, the smelting and synthetic-rutile subprocesses will handle increasing proportions of lower-TiO_2 oxide feeds and will allow such material to cross over into the chloride process.

The meaning of a titanium-mineral resource

Because the economic value of a titanium-mineral deposit depends on complex mineralogic factors, cutoff grades cannot be specified with simple chemical values. Orderly comparison of world resources of titanium minerals (Fantel and others, 1986; Lynd, 1988; Towner and others, 1988) has necessitated a definition of an economic resource of titanium minerals that excludes titanium enrichments that are of no present economic interest. The definition most commonly used at present is the test of economic relevance by Force and Lynd (1984):

Only the titanium oxide minerals rutile and its polymorphs [anatase and brookite], altered ilmenite, ilmenite, and perovskite, which are known or thought to have some economic value, are included. . . .Excluded from resources are titanium minerals of finer grain size than 20 μm (0.02 mm), on the grounds that they cannot presently be separated. Where ilmenite is known to be present as separable grains intergrown with magnetite, resources of the ilmenite are included. Where inseparable intergrowths of magnetite and ilmenite together contain 25 percent or more of TiO_2, resource figures are also included on the grounds that this material could be smelted into high-TiO_2 slag. . . .Our figures include only deposits containing at least 1 percent ilmenite or 0.1 percent rutile or linear combinations thereof in unconsolidated deposits, or 10 percent ilmenite or perovskite or 1 percent rutile in hard rocks. Lower grade resources are included if titanium minerals could be produced as byproducts of other minerals already being mined in the same deposits. . . .

Resource distribution and trade networks

All but about 1 percent of 1987 world titanium-mineral production came from eleven countries (Table 4): Australia, Canada, South Africa, Norway, the United States, Malaysia, the U.S.S.R., Sierra Leone, India, China, and Sri Lanka, in order of total TiO_2 content of concentrates. Brazil will probably join this list in the near future.

Resource distribution is also uneven (Fig. 1; Table 4). Brazil has the largest identified resource in terms of contained TiO_2, followed by Australia. The twelve countries mentioned plus New Zealand dominate the resource picture. Resource figures for the U.S.S.R. and China are not adequately known, and the resources of Madagascar and a few other countries may be far greater than are currently recognized.

Thus a lively trade is inevitable between countries mining

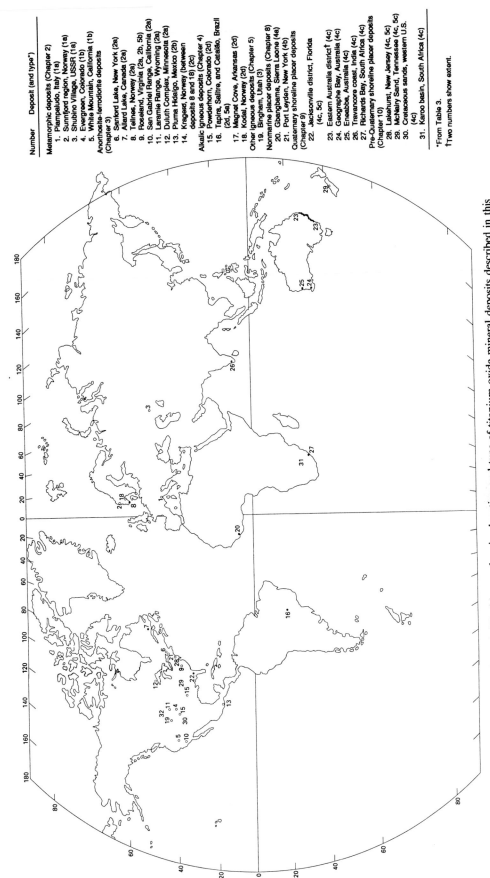

Number	Deposit (and type*)
Metamorphic deposits (Chapter 2)	
1.	Piampaludo, Italy (1a)
2.	Sunnfjord region, Norway (1a)
3.	Shubino Village, USSR (1a)
4.	Evergreen, Colorado (1b)
5.	White Mountain, California (1b)
Anorthosite-ferrodiorite deposits (Chapter 3)	
6.	Sanford Lake, New York (2a)
7.	Allard Lake, Canada (2a)
8.	Tellnes, Norway (2a)
9.	Roseland, Virginia (2a, 2b, 5b)
10.	San Gabriel Range, California (2a)
11.	Laramie Range, Wyoming (2a)
12.	Duluth Complex, Minnesota (2a)
13.	Pluma Hidalgo, Mexico (2b)
14.	Kragerø, Norway (between deposits 8 and 18) (2c)
Alkalic igneous deposits (Chapter 4)	
15.	Powderhorn, Colorado (2d)
16.	Tapira, Salitre, and Catalão, Brazil (2d, 5a)
17.	Magnet Cove, Arkansas (2d)
18.	Kodal, Norway (2d)
Other igneous deposits (Chapter 5)	
19.	Bingham, Utah (3)
Nonmarine placer deposits (Chapter 8)	
20.	Gbangbama, Sierra Leone (4a)
21.	Port Leyden, New York (4b)
Quaternary shoreline placer deposits (Chapter 9)	
22.	Jacksonville district, Florida (4c, 5c)
23.	Eastern Australia district † (4c)
24.	Geographe Bay, Australia (4c)
25.	Eneabba, Australia (4c)
26.	Travancore coast, India (4c)
27.	Richards Bay, South Africa (4c)
Pre-Quaternary shoreline placer deposits (Chapter 10)	
28.	Lakehurst, New Jersey (4c, 5c)
29.	McNairy Sand, Tennessee (4c, 5c)
30.	Cretaceous sands, western U.S. (4c)
31.	Karoo basin, South Africa (4c)

*From Table 3.
†Two numbers show extent.

Figure 1. World map showing location and type of titanium-oxide mineral deposits described in this volume. A more detailed map of U.S. deposits is contained in Force and Lynd (1984). Currently important producers shown solid.

TABLE 5. IDENTIFIED TITANIUM-MINERAL RESOURCES OF THE UNITED STATES, INCLUDING RESERVES*

(In thousand metric tons of contained TiO_2)

State	District or description	Type†	Rutile + polymorphs	Altered ilmenite	Low-TiO_2 ilmenite	Perovskite
Alabama	Sand and gravel	4	....	100	....	....
Arizona	Porphyry copper ore	3	4,000	....	....	....
Arizona	Yuma County	1b	200	....	....	....
Arkansas	Magnet Cove	2d	200	....	....	....
California	San Gabriel Mountains	2a	....	....	4,800	....
California	Ione	4a	....	600	....	....
California	White Mountain	1b	300	....	....	....
Colorado	Powderhorn	2d	....	....	....	20,000
Colorado	Evergreen	1b	200	....	....	....
Florida	Old beach sand	4c(5c)	1,700	9,700	....	....
Florida	Phosphate	4	....	200	....	....
Georgia	Old beach sand	4c	500	2,400	....	....
Georgia	Silica sand	4	100	200	....	....
Idaho	Latah County clay	5	....	....	1,300	....
Maryland	Harford County	1c	700	....	....	....
Minnesota	Duluth Complex	2a	....	....	900	....
Mississippi	Ship Island	4c	....	....	100	....
New Jersey	Lakehurst	4c	....	10,100	....	....
New Mexico	Cretaceous sandstones	4c	....	....	700	....
New York	Sanford Lake	2a	....	....	8,600	....
New York	Port Leyden	4b	....	....	6,300	....
North Carolina	Yadkin Valley	2(a?)	....	....	200	....
North Carolina	Other	4c	....	400	....	....
Oklahoma	Wichita Mountains	4a	....	....	3,900	....
Oregon	Salem bauxite	5	....	....	1,800	....
South Carolina	Hilton Head	4c	100	300	....	....
South Carolina	Charleston	4c	100	1,100	....	....
Tennessee	Cretaceous sand	4c	1,300	8,400	....	....
Utah	Bingham	3	4,000	....	....	....
Virginia	Roseland-Piney River	2a,b(5b)	1,000	....	5,500	....
Virginia	Willis Mountain kyanite	1b	300	....	....	....
Washington	Spokane clays	5	....	....	400	....
Wyoming	Cretaceous sandstones	4c	....	....	500	....
Totals by deposit types		1b	1,000	....	....	....
		1c	700	....	....	....
		2a	....	....	20,000	....
		2b	1,000	....	....	....
		2d	200	....	....	20,000
		3	8,000	....	....	....
		4	100	500	....	....
		4a	....	600	3,900	....
		4b	....	....	6,300	....
		4c	3,700	32,400	1,300	....
		5	....	....	3,500	....
Total by mineral			14,700	33,500	35,000	20,000
Total all minerals	103,200					

*Modified from Force and Lynd (1984); ...., none reported or negligible.

†From Table 3.

titanium minerals and those producing and/or consuming TiO_2 pigments and titanium metal. Prominent among producers having few titanium-mineral resources are Japan, the United Kingdom, and West Germany. The United States is also heavily dependent on imported titanium minerals; although U.S. titanium-mineral resources are considerable (Table 5), they are overshadowed by the country's needs. The United States imports not only titanium minerals but also titanium metal.

Countries that do not produce sufficient amounts of a badly needed mineral may refer to that mineral as strategic. In the United States, titanium minerals are considered among the top ten strategic mineral commodities, partly because of the need for metallic titanium for the aircraft industry. Internationally, titanium is considered sufficiently strategic that it was chosen for study in the International Strategic Minerals Inventory (Towner and others, 1988).

Byproduct-coproduct relations

Titanium minerals are produced as byproducts of a number of other mineral commodities, most notably tin from alluvial mining in southeast Asia. Titanium-mineral resources, not recovered, are present in a large number of other types of active deposits (Table 6). Potentially the most important is rutile from porphyry-type deposits, discussed in Chapter 5.

Conversely, byproducts of titanium-mineral mining are also numerous; these vary considerably among types of deposits (Table 6). Zircon from shoreline-complex placer deposits is worthy of special mention, as it attains coproduct status in some districts.

TABLE 6. BYPRODUCT RELATIONS AMONG TITANIUM MINERALS AND OTHER MINERAL COMMODITIES*

	Currently Recovered	Potentially Recoverable
Titanium as byproduct[†]	Fluvial tin deposits Silica sand deposits (recovered in small part) Palabora-type deposits Nepheline syenite deposits (recovered in USSR only)	Porphyry Cu, Mo deposits Sand and gravel deposits Silica sand deposits Aluminosilicate rock deposits Bauxite (some) deposits Feldspar deposits from anorthosite Detrital phosphorite deposits Sandstone uranium deposits Mafic V, Pt, Cr, Ni-Cu deposits
Titanium as major product[§]	From shoreline placer deposits: Zircon Aluminosilicates Monazite From anorthosite-ferrodiorite deposits: Magnetite Vanadium Apatite	From alkalic deposits: Niobium Rare earths From eclogite deposits: Garnet

*Modified from Force, 1976c, 1980a.
[†]Titanium minerals as byproducts of established mining operations of other commodities.
[§]Other minerals as byproducts of titanium-mineral mining.

Chapter 2.

Occurrence and deposits of titanium oxide minerals in metamorphic rocks

Titanium oxide minerals formed by metamorphic processes play three roles in the economic geology of titanium. First, some extreme metamorphic processes actually form deposits of economic interest. Such deposits account for less than 2 percent of the identified resources of the United States, but the eclogite type of metamorphic deposit may become more important elsewhere in the future. Second, titanium oxide minerals from metamorphic rocks are the predominant source of those minerals in placer deposits. Third, the massif anorthosite-ferrodiorite suite of igneous rocks that contain important titanium-mineral resources is characteristically emplaced under particular metamorphic conditions. These igneous rocks in turn form a subordinate source of detrital titanium oxide minerals in placer deposits. Because of these roles of metamorphism in influencing igneous deposits and controlling placer deposits, titanium will be traced through metamorphic processes with some diligence.

Titanium contents of rocks generally remain approximately constant during metamorphism. In fact, titanium is so unusually immobile in many subsolidus geologic processes that calculations of the amount of change in other elements are commonly normalized to titanium, assuming it to be conserved. The partitioning of titanium between oxide and silicate phases, however, varies markedly among metamorphic facies. For example, 0.6 percent TiO_2 in some ordinary metamorphic rocks (Table 1) may be present as 1.5 percent sphene at low metamorphic grades or as 1.2 percent ilmenite or 0.6 percent rutile at high grades.

VARIATION IN PARTITIONING WITH ROCK COMPOSITION

Titanium partitioning between oxide and silicate phases is also a function of rock composition. The compositional variables of greatest importance appear to be the Al/Ca ratio, which governs reactions of the type:

$$\text{sphene} + Al_2O_3 = \text{rutile} + \text{plagioclase} \qquad (2\text{-}1)$$

and the Fe/Mg ratio, which governs reactions of the type:

$$\text{pyroxene} + \text{rutile} = \text{ilmenite} + \text{quartz} + MgO. \qquad (2\text{-}2)$$

Figure 2 shows the relation of these variables and of metamorphic grade to titanium mineralogy.

Volatile constituents may affect partitioning of titanium among oxide minerals and between oxides and silicates. Volatiles of importance include oxygen, carbon dioxide, and sulfur. Oxygen controls reactions of the type:

$$\text{ilmenite} + O_2 = \text{magnetite} + \text{rutile} \qquad (2\text{-}3)$$

and sulfur controls reactions of the type:

$$\text{ilmenite} + S_2 = \text{pyrite} + \text{rutile}. \qquad (2\text{-}4)$$

Similar reactions can also be written to show liberation of titanium from silicates. CO_2 enters into the reaction:

$$\text{sphene} + CO_2 = \text{rutile} + \text{calcite} + \text{quartz}. \qquad (2\text{-}5)$$

VARIATION IN PARTITIONING WITH METAMORPHIC GRADE

For convenience of discussion, metamorphism is divided into higher and lower grades between the upper and lower amphibolite facies. For rocks of high-pressure facies series, it is divided between the higher and lower blueschist facies.

Lower-grade metamorphic rocks

Most rocks at lower metamorphic grades contain the bulk of their titanium in silicates. Sphene is the most important carrier of titanium in many such rocks, and biotite and hornblende are also important carriers in some rocks. Figure 2 shows the relation of sphene stability to rock composition and metamorphic grade.

11

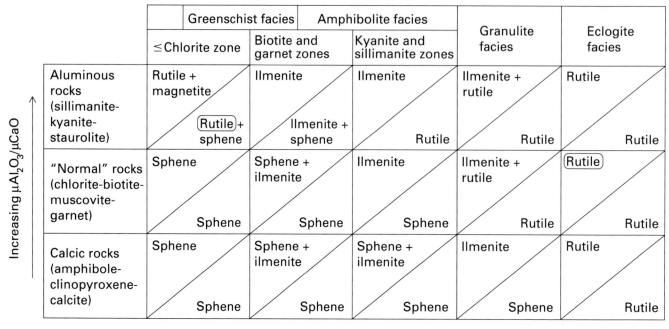

Figure 2. Schematic relation of rutile, ilmenite, and sphene occurrence to composition and grade in metamorphic rocks. Circled minerals indicate deposits, currently subeconomic; μ is chemical potential.

In subgreenschist and lower greenschist (chlorite zone) facies, the predominant carrier of titanium in most rocks is apparently sphene (Force, 1976b). Sphene as fine dusty aggregates is a common and moderately abundant phase in these rocks, both felsic and mafic. Relict titanium oxide minerals commonly are in advanced stages of recrystallization; for example, ilmenite and rutile in placer concentrations metamorphosed at low grades are commonly recrystallized to sphene and magnetite (Goldsmith and Force, 1978, Fig. 3). Magnetite at this grade has low TiO_2 contents (Abdullah and Atherton, 1964), and there is little titanium in chlorite. Extremely fine-grained anatase may be intergrown with sphene in some rocks (Rumble, 1976; Herz and Force, 1987). Rocks having very high Al/Ca ratios may contain fine rutile (cf., Zen, 1960). Those with high Al/Ca ratios, high Fe/Mg ratios, and high oxygen fugacity may show rutile + magnetite or hematite assemblages (Meilke and Schreyer, 1972; Rumble, 1973, 1976).

In upper greenschist (biotite zone)-facies rocks, biotite becomes a significant carrier of titanium. At this grade, biotite typically contains a little more than 1 percent TiO_2. Many rocks, especially mafic ones, continue to contain a considerable share of their titanium as sphene. Ilmenite may form in place of rutile in the high-oxygen pelitic rocks by reaction (2-3).

In the lower amphibolite facies, hornblende becomes a carrier of titanium in mafic rocks. Hornblende at this metamorphic grade contains a little more than 1 percent TiO_2. Sphene continues as an important carrier of titanium in rocks of a wide range of compositions and becomes coarser grained. Ilmenite coexisting with hematite is also common (Rumble, 1976; Nedelcu, 1986); it may form with plagioclase by the breakdown of sphene and muscovite.

Retrograde reactions in originally high-grade rocks are, of course, similar to reactions in low-grade rocks and can commonly be recognized by rims of sphene and biotite around titanium oxide minerals. Chemical microenvironments limited to single grains can be recognized in some retrograded rock textures; for example, rutile needles in chlorite may replace titaniferous biotite grains.

Rutile in altered lower-grade metamorphic rocks

Metasomatic additions of constituents such as magnesium or sulfur, or depletion in calcium, may permit the formation of rutile (equations [2-1] to [2-4]). The introduction of magnesium into "blackwall" metasomatic zones adjacent to serpentinites ties up available iron; rutile is therefore commonly found in such blackwalls (Chidester, 1962). Southwick (1968) and Herz and Valentine (1970) have described an unusual chlorite rock containing coarse rutile, magnetite, and apatite that apparently formed in this manner; it is discussed later in this chapter as a resource.

In a similar manner, rutile is found in the wall rocks of metamorphosed massive sulfide deposits. Nesbitt and Kelly (1980) show that around the Ducktown sulfide bodies of eastern Tennessee, metamorphosed to staurolite grade, ilmenite in wall rocks has responded to local high sulfur pressure in going to pyrrhotite and rutile. However, the economic importance of this occurrence type is thus far limited to prospecting for massive sulfides.

Hydrothermal stripping of cations from metavolcanic rocks and formation of pyrite have led to appreciable rutile contents along with andalusite and topaz in several pyrophyllite deposits of the slate belt in North Carolina and South Carolina (Schmidt, 1985; oral communication, 1986). These deposits occur in host rocks representing the greenschist facies (even though kyanite is locally present; Carpenter, 1982). Schmidt (1985) and Carpenter and Allard (1982) reviewed a large number of related world occurrences of aluminous hydrothermal systems with associated rutile in metavolcanic rocks of low metamorphic grades. Rutile in aluminous hydrothermal rocks persists to higher-grade metamorphic assemblages (Geijer, 1964; Marsh and Sheridan, 1976). Some economically interesting rutile occurrences are described later in this chapter.

High-grade metamorphic rocks of "normal" P-T ratios

Ramberg (1948, 1952) was the first to show that titanium is transferred from titanium-bearing silicates at lower metamorphic grades into titanium oxide minerals at high grades. This process was further documented by Force (1976b) and Goldsmith and Force (1978). Figure 2 shows the influence of bulk composition on resulting mineralogy.

In the upper amphibolite facies, rutile becomes common in some pelitic rocks, and ilmenite is common over a wide range of lithologies. These phases form partly at the expense of sphene, which becomes less common in these rocks as the calcium content of plagioclase increases (as in reaction [2-1]). Spear (1981) found that a similar reaction in amphibolites consumes sphene to form ilmenite, with some compositional change of plagioclases and amphiboles; this reaction is favored by high oxygen fugacity as well as increasing temperature.

Goldsmith and Force (1978) found that rutile occurs in pelitic rocks at grades as low as the kyanite zone in units in which the Al/Ca ratio is greater than 50. Wall rocks having lower Al/Ca ratios contain sphene instead. Textural relations suggest that rutile forms with garnet at the expense of ilmenite, possibly by the reaction:

ilmenite + muscovite + quartz = almandine garnet + rutile + biotite
(2-6)

(see also Ghent and Stout, 1984, for related reactions). At higher metamorphic grades, an increasing variety of rocks contain rutile (Fig. 3).

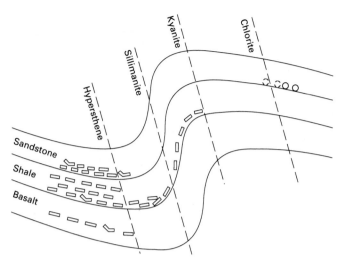

Figure 3. Diagram of rutile distribution in some common metamorphic rocks (from Goldsmith and Force, 1978, Figs. 5 and 6; Force, 1980b, Fig. 1). Rectangles are metamorphic rutile (some twinned); circles are detrital rutile, dashed where retrograding.

Rutile may form in sulfide-bearing rocks by prograde conversion of pyrite to pyrrhotite:

pyrite + ilmenite = pyrrhotite + rutile. (2-7)

Such reactions involving sulfides may also liberate titanium from titaniferous biotite and hornblende by forming more magnesian silicates plus oxide minerals, including rutile (Robinson and Tracy, 1977; Robinson and others, 1982; Mohr and Newton, 1983).

In the granulite facies, titanium oxide minerals become the dominant carriers of titanium in most lithologies. The granulite facies by definition involves formation of pyroxenes at the expense of biotite and hornblende; this leads to titanium liberation, as the pyroxene contains much less titanium than biotite and hornblende. The approximate reactions are:

biotite + quartz = orthopyroxene + garnet + orthoclase + rutile + vapor
(2-8)
and:

hornblende + quartz = orthopyroxene + plagioclase + ilmenite + vapor.
(2-9)

With increasing metamorphic grade, biotite and hornblende accommodate increasing amounts of TiO_2, until in the granulite facies, biotite may contain up to 6 percent TiO_2 and hornblende up to 4 percent TiO_2 (reviewed by Force, 1976b; Guidotti and others, 1977; Spear, 1981; Dymek, 1983). The TiO_2 contents of these minerals are a function of Fe/Mg ratios as well as metamorphic grade (Guidotti and others, 1977; Robinson and others, 1982). However, the proportion of titanium present in biotite and

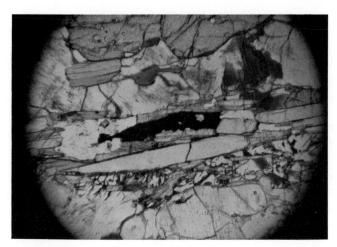

Figure 4. Photomicrograph of rutile (dark) in graphite-sillimanite-biotite-garnet-K-spar gneiss, Franklin, North Carolina. Transmitted plane light, 6-mm field.

hornblende in a rock *decreases* with increasing grade, because of reactions consuming those minerals (Force, 1976a, Table 6).

Sphene becomes uncommon in the granulite facies except in the most calcic rocks (Ramberg, 1952; Turner, 1968) because of reactions such as:

$$\text{sphene} + \text{hornblende} + \text{sodic plagioclase} =$$
$$\text{clinopyroxene} + \text{calcic plagioclase} + \text{ilmenite.} \qquad (2\text{-}10)$$

Magnetite also commonly becomes less abundant in this facies, typically shown by lessened gradients on aeromagnetic maps. Titanium liberated from biotite, hornblende, sphene, and magnetite goes into ilmenite and rutile, resulting in the greatest partitioning toward titanium-oxide minerals possible for gneisses, schists, and amphibolites (Table 1).

In these granulite terranes, relatively mafic lithologies typically contain ilmenite, and leucocratic lithologies contain rutile (Figures 3, 4). In some situations, especially in the pyroxene granulite subfacies, rocks of intermediate composition contain rutile also. Both rutile and ilmenite in such rocks are typically coarse and locally are free of intergrowths (Herz and Force, 1987), but retrogression may produce rims of other minerals.

Eclogites are locally the end members of high-grade metamorphism of "normal" facies series (Coleman and others, 1965). These rocks, which contain rutile as a characteristic phase and little ilmenite, are described later in this chapter.

High-grade metamorphic rocks of high-pressure type

Metamorphic rocks formed at low ratios of temperature to pressure form a separate metamorphic facies series. At low metamorphic grades, these rocks are similar in titanium mineralogy to normal low-grade metamorphic rocks (Ernst and others, 1970; Seki and others, 1971; Itaya and Otsuki, 1978); that

is, sphene and other silicates carry most of the titanium, ilmenite is locally common, and rutile is present in some rocks of unusual composition. In higher-grade rocks of this type, rutile becomes common (Blake and Morgan, 1976; Itaya and Banno, 1980); the high-grade facies was named the epidote-rutile blueschist facies by Taylor and Coleman (1968). In these high-grade rocks, rutile typically occurs with garnet, sodic and calcic amphiboles, and epidote. Ilmenite is not commonly reported, and sphene is present as a retrograde mineral. Blake and Morgan describe occurrences of such rutile-bearing rocks from California, New Caledonia, Japan, the European Alps, and Venezuela.

Rutile is a characteristic phase of eclogitic rocks, the highest temperature and pressure member of this series. Again, ilmenite is commonly not reported, and sphene is common only as a retrograde mineral. As other titaniferous silicates are absent in eclogite, virtually all TiO_2 in the rock may be present as rutile. Where eclogite has a ferrogabbroic composition, rutile contents may be more than 5 percent (Chesnokov, 1960; Cortesogno and others, 1977). Such eclogites are of economic interest, and three districts are described in the next section.

In many occurrences, these high-grade metamorphic rocks occur as tectonic inclusions in lower-grade terranes. However, in New Caledonia (Black, 1977), a transition from lawsonitic schists to eclogitic gneisses is exposed, with rutile occurrences limited to the high-grade epidote zone. Other relatively complete transitions are known from Ecuador, Venezuela, Norway, Italy, and Japan.

MAJOR DEPOSITS

At present, no titanium minerals are produced from metamorphic deposits. The eclogite-type deposit has the greatest economic potential, and such deposits are therefore emphasized as the first three descriptions below. The different types of deposits are put in their metamorphic context in Figure 2.

Rutile is the economic mineral in metamorphic titanium-mineral deposits. Metamorphism does not enrich rocks in titanium; thus, available titanium in a rock must be present in a most valuable form in order for the rock to qualify as a resource.

Piampaludo deposit, Italy

Development work since 1974 has been conducted at Piampaludo, mostly by Geomineraria Italiana, on a single large tectonic inclusion of eclogite in serpentine in the Ligurian Alps (Fig. 1) of northern Italy (Mancini and others, 1979; Clerici and others, 1981). Similar but smaller bodies are abundant in the region (Martinis and Pasquare, 1971; Cortesogno and others, 1977) and commonly have a ferrogabbroic composition (4.6 percent TiO_2, 18.2 percent total Fe oxides).

The Piampaludo body is exposed in several hills and the intervening gorges of the Orba and Orbarina Rivers (Fig. 5), in mountainous country dotted with attractive villages. Its outcrop area is about $500 \times 1,800$ m.

Along the Orba River section through the potential Piampa-

Figure 5. Annotated photograph of the Piampaludo, Italy, deposit from the Orba River gorge. E, eclogite.

2.7 to 9.3 percent rutile. Rutile that is probably too fine to be separated (<50 μm) was not counted. Correcting for density, rutile averages 6.2 percent by weight, a figure consistent with chemical analyses of the body (6 percent TiO_2) by Mancini and others (1979). Modal analyses of eclogites of the entire district average 5.8 volume percent rutile (Cortesogno and others, 1977).

The composition of rutile from Piampaludo has not yet been reported. Elaine McGee (written communication, 1986) found by electron microprobe that the chemical impurities Nb_2O_5, Cr_2O_3, MgO, MnO, FeO, CaO, and Al_2O_3 in this rutile constituted less than 1 percent each and that TiO_2 content was over 95 percent.

Clerici and others (1981) report that rutile liberation from its eclogite host during crushing is adequate. A good rutile concentrate was made with adequate recovery.

Mancini and others (1979) list proven ore of the Piampaludo deposit as 150 million metric tons, based on surface lithology and nine drill holes of 100 m depth. Their estimate of probable and possible ore, an additional 700 million metric tons, is optimistic in my opinion. At the northern end of the body, where no drill holes were sited, the basal contact and foliation in both host and eclogite appear to dip gently south. Thus, the base of the body may be shallow there. The amount of proven ore, however, implies 9.3 million metric tons of rutile, potentially making Piampaludo a deposit of world importance.

Sunnfjord region, Norway

Eclogite is present as discrete bodies of various sizes forming a subordinate constituent of gneiss terrane throughout the mountainous Western Gneiss Region of Norway. The origin of these eclogites has been debated since the pioneering work of Eskola (1921). Recent reviews are by Griffin and Mork (1981), Griffin and others (1985), and Griffin (1987).

Formation of this terrane is complex, as one might imagine. Mafic rocks of several origins apparently were metamorphosed to eclogite during the early Paleozoic. Among the precursors were mafic portions of a layered ultramafic-anorthositic-gabbroic-mangeritic complex possibly emplaced in supracrustal rocks. Other precursors were mafic dikes and sills. Metamorphism of the entire terrane under eclogite-facies conditions preceded preferential retrogression of felsic and supracrustal host rocks to granulite and amphibolite facies. In addition, dismemberment into isolated bodies occurred by boudinage of competent mafic rocks and by injection of granitoid melts, either before or after eclogite-facies metamorphism. Commonly, eclogite bodies are present as horizons or even as thin cumulate layers, in bodies that also contain garnet peridotite, anorthosite and troctolite, mangerite, and/or oxide-rich bands. The margins of bodies of eclogite are commonly retrograded to amphibolite.

Eclogites of ferrogabbroic composition are known as large bodies only in the Sunnfjord region (Fig. 1). The eclogite bodies are limited mostly to one particular unit in the gneiss terrane (Skjerlie and Pringle, 1978; mapped by Kildal, 1970). In this

ludo orebody, eclogite is dominantly unsheared and unaltered. Serpentine partings are rare, and the eclogite is otherwise quite homogeneous. A strong planar fabric dipping south is pre- or synmetamorphic.

Ore in thin section shows porphyroblastic garnet and pyroxene megacrysts in a mass of blue to green amphibole, much of it foliated, along with epidote, micas, and rutile (Fig. 6). Rutile occurs in aggregates strung out parallel to foliation. The aggregates commonly measure 1 to 2 mm wide and contain 40 to 90 percent rutile, mostly as crystals of 100 μm or more minimum diameter (Fig. 6). Other minerals in the aggregates are amphibole, epidote, ilmenite, and minor garnet. Alteration rims of sphene around rutile are minor in most specimens.

The literature reports the rutile contents of Piampaludo ore only by chemical analysis for TiO_2 (Mancini and others, 1979), an ambiguous procedure in the presence of sphene. Based on my point counts of five typical, well-distributed specimens, rutile averages 5.3 percent by volume. Individual specimens range from region, the eclogite bodies are essentially large xenoliths, as most of the unsheared eclogite margins are against granitoid rocks containing abundant small retrograded mafic xenoliths (Naustdal southern contact and Kvineset lens 2, Krogh, 1980, Fig. 2; Engebøfjellet southern contact, my observation).

The less retrograded portions of these ferrogabbroic eclogite bodies contain as much as 6 percent rutile (by weight). Rutile resources of the district have been evaluated by Korneliussen and Foslie (1986). Detailed maps are contained in Korneliussen (1980). Three of the larger bodies, Naustdal, Engebøfjellet, and Fureviknipa, contain at least 100 million metric tons of eclogite each and average 2.7 to 3.1 percent rutile. Part of the Naustdal body is the site of a town.

Rutile occurs as grains averaging 0.1 to 0.2 mm in diameter, commonly in aggregates, among megacrystic omphacite, euhedral garnet, and minor interstitial amphibole (Fig. 7). Microprobe analyses by Korneliussen and Foslie (1986) show that rutile contains 98.9 to 99.4 percent TiO_2, with vanadium the primary contaminant. Ilmenite is present with rutile as fine intergrowths throughout the bodies and as rims on rutile aggregates in retrograded zones irregularly distributed through the bodies.

Shubino Village deposit, U.S.S.R.

Eclogites and related rutile-bearing rocks (Chesnokov, 1960) occur in schist terrane of the southern Ural Mountains near Shubino Village (Fig. 1). The eclogites are of ferrogabbroic composition and form bodies up to 200 × 1,000 m, elongated conformable to foliation of the schist matrix. Retrograding along the margins of these bodies is advanced, but in their interiors, rutile is said by Chesnokov to be fresh and its distribution homogeneous. Such zones are up to 60 m thick. They apparently have been mined for rutile (Blake and Morgan, 1976).

Rutile is present mostly in the groundmass. Median grain size of the rutile is 0.12 mm (Chesnokov, 1960, Fig. 7). The

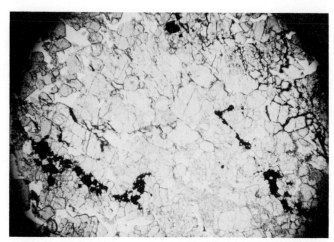

Figure 7. Photomicrograph of rutile-bearing eclogite from Engebøfjellet, Norway, showing rutile (dark), omphacite megacrysts, garnet, and matrix amphibole and quartz. Transmitted plane light, 6-mm field.

average rutile content of the bodies is not given, but TiO_2 contents range from 3.3 to 5.35 percent. Some of this TiO_2, however, is present as small rutile inclusions in garnet and as sphene.

Chemical analysis of rutile from this eclogite shows 94.2 percent TiO_2, with Fe_2O_3 and Al_2O_3 the principal impurities (Chesnokov, 1960, Table 7). No analyses are listed for several elemental contaminants of interest.

Summary of eclogite-type deposits

Some features shown in common by the three known rutile deposits in eclogite are ferrogabbroic parent with consequent TiO_2 enrichment relative to normal eclogite; unaltered, unsheared eclogite with relatively homogeneous rutile distribution; dimensions from 0.1 to 4 km^2; and rutile grain size averaging 0.1 mm or greater. Rutile grades are 3 percent or greater, and rutile tonnages range from about 0.1 to 10 million metric tons per body. Host terranes of these eclogite bodies vary widely. Eclogites of ferrogabbroic composition appear to occur in swarms that may not include eclogites having more normal compositions.

Garnet composition, as well as rutile abundance, shows the effect of ferrogabbroic composition in all three rutile deposits. The garnet is almandine (Chesnokov, 1960; Binns, 1967; Ernst, 1976), whereas more pyropic garnet is normally characteristic of eclogite. The omphacitic composition of pyroxene is unaffected.

Aluminosilicate-rutile deposits

Probably the second most important type of metamorphic titanium-mineral deposit is the aluminosilicate type. These deposits also contain rutile as the economic titanium mineral. The deposit type has been reviewed by Geijer (1964), Wise (1975), Marsh and Sheridan (1976), and Schmidt (1985).

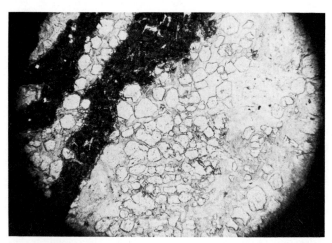

Figure 6. Photomicrograph of Piampaludo rutile deposit showing foliated aggregate of rutile crystals (dark), euhedral garnet, and omphacite megacryst (upper right). Transmitted plane light, 2-mm field.

These deposits apparently form mostly from volcanogenic parent rocks by premetamorphic to synmetamorphic hydrothermal stripping of some chemical constituents, especially alkalies. Removal of calcium and iron is most pertinent to the formation of rutile. Titanium and aluminum are left behind, residually enriched by the removal of the other constituents. Fluorine and perhaps phosphorus may be added. Rutile crystallizes largely because of the paucity of chemical constituents that stabilize competing titanium minerals.

The mineralogy of these deposits is quite distinctive, varying largely with metamorphic grade. Aluminosilicate minerals are abundant and may dominate over quartz; these may vary, however, from sillimanite at higher metamorphic grades through kyanite and andalusite to pyrophyllite at lower grades. Topaz (aluminum fluorine silicate) and aluminous phosphates such as lazulite are common. Pyrite is locally abundant.

Chemical stripping at constant volume has resulted in open space in some deposits of this type. Rutile and other minerals may thus be euhedral. Some of the more famous rutile mineral localities, such as Graves Mountain, Georgia, occur in deposits of this type.

Rutile grade in these deposits is commonly about 1 percent or less, and tonnages are commonly small. Thus they are promising as resources only where the other constituents of the rock can be mined as primary products or where the rock is locally enriched in TiO_2. Both cases do occur.

The kyanite, pyrophyllite, and other aluminosilicate deposits of the southeastern United States are examples of deposits in which rutile is a possible byproduct. Figures in Espenshade and Potter (1960) imply rutile resources of about 300,000 metric tons in active and marginal kyanite deposits. Rutile grade averages about 0.4 percent, and average grain size appears to be less than 0.1 mm.

The pyrophyllite-topaz-andalusite rocks of the same region, mined for pyrophyllite, also contain rutile as a characteristic accessory mineral (Schmidt, 1985), mostly as aggregates of rutile crystals only rarely coarser than 20×50 μm (Schmidt, oral communication, 1986). No rutile resource calculations for this type of occurrence have been attempted. Two deposits of this type having greater economic potential conferred by greater TiO_2 contents are the Evergreen deposit of Colorado and the White Mountain (or Champion) deposit of California.

The Evergreen deposit, in the Colorado Front Range, is a sillimanite-quartz-topaz gneiss forming a deformed stratiform unit up to 30 m thick in high-grade gneiss terrane (Sheridan and others, 1968; Marsh and Sheridan, 1976). Some sections through the unit contain 2 to 5 percent rutile. Rutile is equant to prismatic and commonly is 0.5 mm in lesser diameter. Rutile is fairly free of contaminants other than iron and calcium, and its TiO_2 content is everywhere greater than 98 percent (Marsh and Sheridan, 1976, Table 2). Resources listed by Force and Lynd (1984) are 200,000 metric tons of rutile. Residential land use prevents the mining of this deposit.

The White Mountain deposit of California, formerly mined

Figure 8. Annotated photograph of the White Mountain (Champion) deposit, California, after Marsh (1979). Sch, schist; qmp, quartz monzonite porphyry; ba, bleached argillic rock; qtar, quartz-topaz-andalusite-rutile rock.

for andalusite, is near Bishop, east of the Sierra Nevada (Gross and Parwell, 1969; Wise, 1977; mapped by Crowder and Sheridan, 1972). It is resistant andalusite-topaz-quartz rock (Fig. 8) that apparently formed by hydrothermal leaching of greenschist-facies volcanogenic rocks. Rutile is most commonly present in concentrations of less than 1 percent by weight, but it is present in concentrations averaging about 2 percent in an area $60 \times 1,300$ m (reconnaissance observations in 1981 of Force and Marsh). On this basis, Force and Lynd (1984) list a rutile resource of 300,000 metric tons. Rutile grain size is highly variable, from 0.01 to 0.2 mm. The most economically attractive lithology is an atypical saccharoidal breccia with about 4 percent rutile confined to the pink matrix (Fig. 9).

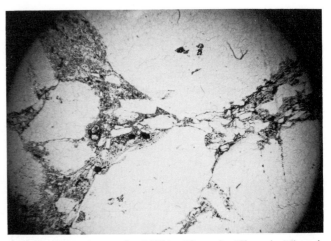

Figure 9. Photomicrograph of White Mountain (Champion) breccia with topaz-rich fragments. The matrix contains more than 10 percent rutile, in grain sizes from 10 to 100 microns. Transmitted plane light, 6-mm field.

Metamorphic-metasomatic blackwalls

The Dinning prospect of Harford County, Maryland, is the only blackwall-type deposit thought to have economic potential. A chlorite rock there separates a serpentinite body from greenschist-facies schist and contains coarse rutile, apatite, and magnetite (Southwick, 1968). Average rutile content is about 1 percent but ranges as high as 4 percent (Herz and Valentine, 1970). Force and Lynd (1984) list a rutile resource of 700,000 metric tons. The rutile contains appreciable Fe, Si, Al, and Mg. The high Fe, Ti, and P contents of this chlorite rock suggest that the parent rock was a ferrogabbro.

ECONOMIC PROGNOSIS

The eclogite type of rutile deposit can yield hundreds of millions of tons of ore containing up to about 6 percent rutile. The rutile reportedly has a composition that would allow chloride processing, and it is not too difficult to separate. In the current state of change in the world supply of titanium minerals, eclogite-type rutile deposits could emerge as a major source of titanium minerals.

The other types of rutile deposits in metamorphic rocks tend to be either low grade or small. Not even the described examples contain as much as a million tons of rutile. Probably such deposits could not compete in current world markets.

METHODS OF EXPLORATION

Titanium oxide minerals in metamorphic rocks are the primary source of those minerals in placer deposits (cf., Force, 1980b). Thus, knowledge of metamorphic terranes can be used in placer exploration. Placer deposits and their exploration are discussed in Chapters 7 through 10.

Eclogite-type deposits are apparently the only metamorphic deposit type of potential world importance. The smaller aluminosilicate and blackwall deposits probably warrant exploration only for another commodity as primary product, and warrant evaluation as rutile deposits only where they are already located. The following paragraphs propose an exploration method for rutile in eclogite-type deposits; probably no such exploration has as yet been attempted.

Eclogite-type deposits tend to occur in different lithotectonic terranes. The explorationist is thus forced to begin with a terrane in which eclogite bodies are known to occur. The first goal in exploring for a rutile deposit is to find a sizable body of un-sheared, unaltered eclogite of ferrogabbroic composition. In the case of districts described in this chapter, locating such eclogites could have been done to a great extent with existing geologic maps and existing lithologic and chemical descriptions. This is probably true in many other terranes also, because intellectual curiosity about eclogites far predated any economic interest in their rutile contents. It appears that eclogite bodies of ferrogabbroic compositions occur in swarms, and these certainly offer the best hunting.

Where existing information about eclogite distribution is scanty, the simplest exploration procedure probably is alluvial prospecting for rutile in nonglaciated terranes known to contain eclogite bodies, followed by physical inspection of eclogite bodies for outcrop area, degree of alteration and shearing, and rutile content and grain size via thin section. Determination of garnet composition may be a guide, either in alluvial or bedrock samples, but in an unexpected way; the ferrogabbroic eclogites that contain economically interesting rutile contents contain almandine rather than pyropic garnet. Clearly, exploration with pyrope will at best find lean rutile in eclogite. In terranes where garnet is all contained in eclogite bodies, the presence of almandine in alluvial samples may point to valuable rutile deposits.

Chapter 3.

Titanium oxide minerals in anorthosite-ferrodiorite massifs

Only two suites of igneous rocks contain titanium-mineral deposits that are currently considered valuable: the anorthosite-ferrodiorite massifs discussed in this chapter and alkalic ring complexes discussed in Chapter 4. The reason that igneous deposits are restricted to these two suites lies in two factors: (1) the titanium in these rocks is predominantly in titanium oxide minerals, and (2) these rocks include facies that have high TiO_2 contents (Table 1). The behavior of titanium in other igneous rock clans is discussed in Chapter 5.

The role of metamorphism discussed in the previous chapter extends, in my opinion, to the anorthosite-ferrodiorite igneous massifs. These rocks are characteristically found in high-grade metamorphic terranes (Anderson and Morin, 1969; Ashwal, 1982b).[1] In many districts, such as Sanford Lake, New York, and Roseland, Virginia, there is adequate evidence that emplacement was during and/or before metamorphism, that is, some high-grade metamorphism and attendant deformation affected the igneous rocks. Thus the equilibria governing the partitioning of titanium between silicates and oxides in metamorphic rocks also apply to this clan of igneous rocks. This chapter explores a transition to similar igneous assemblages in more stratiform bodies emplaced in lower-grade metamorphic terranes.

The anorthosite massifs occurring in metamorphic terranes of high grade were emplaced partly by igneous processes and partly by tectonic or diapiric processes and are deformed, commonly into domes. The massifs are accompanied by ferrodiorites, ferrogabbros, charnockites, and rapakivi granites (Emslie, 1978). The ferrodiorites and gabbros are typically younger than anorthosite, and they show distinct geochemical relations (Ashwal, 1982b), but their association with anorthosite is so strong that there must be a cogenetic relation. The reported range in age of anorthosite massifs is rather narrow, from 1700 to 900 m.y. old (cf., Anderson and Morin, 1969).

The massif anorthosites vary in composition from andesine-type, locally antiperthitic, to labradorite-type, and some have marginal facies containing mafic minerals. Plagioclase megacrysts are commonly separated by a finer matrix representing megacryst granulation and/or igneous groundmass. Dikes of anorthosite in country rock are common only in the more sodium- and potassium-rich bodies. True anorthosites have very low TiO_2 contents (Table 1) and normally contain few titanium oxide minerals.

In the anorthosite-ferrodiorite massifs, it is the ferrodiorites and related rocks that are enriched in TiO_2 and contain titanium oxide minerals in abundance. These bodies of ferrodiorite to ferrogabbro contain significantly more TiO_2 and P_2O_5 than do other diorites and gabbros. Their total iron oxide content is also unusually high, especially relative to MgO (Emslie, 1978). These bodies form igneous sheets structurally overlying the older anorthosites, and they commonly send dikes into anorthosite and other country rocks.

The petrogenesis of this igneous suite is not discussed here, except for those aspects bearing directly on titanium oxide minerals. This topic is complex and is the subject of a large and contentious literature.

Titanium-bearing oxide minerals present in this igneous suite include ilmenite, magnetite, hematite, ulvospinel, and minor rutile. Intergrowths among these minerals are common. The presence or absence of ilmenite as separable single crystals relatively free of intergrowths is a key feature determining the economic value of a deposit. In general, TiO_2 present in magnetite solid solution or as fine intergrowths in magnetite is valueless.

Deposits related to anorthosite-ferrodiorite massifs are discussed in this one chapter, though they are of two distinct types. In one type, the deposits are true igneous rocks formed from titanium-rich liquids. In the other, high-temperature metasomatism between igneous rocks and titanium-bearing wall rocks formed the deposits. The geologic setting of the two types of deposit is similar; indeed, at Roseland the two types are present in the same district but did not form at the same time.

---

[1]For a contrary opinion emphasizing more calcic anorthosites, see Morse (1982).

MAGMATIC ILMENITE DEPOSITS

Ilmenite deposits of magmatic origin are currently the most important type of titanium-mineral deposits in igneous rocks. They constitute approximately 30 percent of both world titanium-mineral reserves and of current production (from data in Lynd, 1988).

Facies

In detail, magmatic ilmenite deposits differ considerably from one another and show great variation internally. The variation can perhaps be best understood if the deposits are considered to occupy positions in a polygon or matrix, with end members reflecting three types of facies variation, each of which varies somewhat independently (Table 7): (1) andesine versus labradorite anorthosite association, (2) ferrodioritic versus anorthositic deposit host, and (3) nelsonitic versus massive oxide facies. There is also a transition to stratiform hosts, which is discussed separately; the description here is restricted to anorthosite-ferrodiorite massifs containing massive to layered rocks rich in oxides, predominantly ilmenite. Within the framework of facies variations, these deposits show remarkable similarities.

Andesine versus labradorite anorthosite association. Massif anorthosites show a spectrum from bodies of labradorite composition to bodies of andesine composition (Anderson and Morin, 1969). Andesine in the latter commonly is antiperthitic and in some bodies contains as much as 4 percent K_2O.

The transition in anorthosite composition corresponds to sympathetic transitions in the compositions of associated mafic rocks and to the nature of associated iron-titanium oxides. Labradorite anorthosite massifs are associated with gabbros and with oxide-rich rocks containing magnetite, ulvospinel, and ilmenite. These phases are commonly too finely intergrown for effective separation. Andesine anorthosite massifs are associated with ferrodiorites and with oxide-rich rocks containing magnetite, hematite, and ilmenite, with only hematite and ilmenite finely intergrown ("hemoilmenite"). Andesine anorthosites rich in K_2O (alkalic andesine anorthosites of Herz, 1969) are accompanied by ferrodiorites rich in K_2O and SiO_2, which in turn are associated with oxide-rich rocks containing nearly stoichiometric intergrowth-free ilmenite or ilmenite intergrown with hematite but largely free of magnetite.

Ferrodioritic versus anorthositic deposit host. In each district described in this chapter, the ilmenite-rich rocks cut anorthosite; in younger, more mafic rocks, however, ilmenite-rich rocks parallel igneous layering. In most districts, both types of ilmenite deposit occur in adjacent rock units. It is here proposed that these deposit types are correlative facies, both postanorthosite but coeval with ferrodiorite or gabbro.

In the anorthositic facies, ilmenite-rich rocks form unambiguous dike-like bodies of variable size, commonly with apophyses. Contacts are sharp, though they may be modified by metamorphic recrystallization. Anorthosite functions merely as country

TABLE 7. A CLASSIFICATION FOR THE DESCRIBED DEPOSITS OF THE ANORTHOSITE-FERRODIORITE TYPE*

Magmatic ilmenite deposits		Contact-metasomatic rutile deposits	
Massif deposits	Hybrid massif-stratiform deposits	Anorthositic	Albititic
Sanford Lake district (a+l, an+fg, m+n)	Duluth Complex (l, fg, m+n)	Roseland (part) Pluma Hidalgo	Kragerø Beaver Creek
	San Gabriel Range (a, an+fg, n)		
Allard Lake district (a, an+fg, m)			
Tellnes (a, an, m)			
Roseland (part) (aa, an+fg, n)			
Laramie Range (l, an, m+n)			

*Facies shown in parentheses. Abbreviations of facies: Anorthosite association, l = labradorite, a = andesine, aa = alkalic andesine. Deposit host, an = anorthosite, fg = ferrodiorite-gabbro. Mineralogic facies, n = nelsonitic, m = massive-oxide

rock, and locally the dikes may cut other units, such as gabbro or preanorthosite gneiss.

In the ferrodiorite-gabbro facies, ilmenite-rich lithologies characteristically form layers as thin as fractions of a centimeter. These may define the layering in ferrodiorite or may be parallel to other layering and are concordant to the base of the body. Iron-titanium oxide minerals and apatite together form a net-vein system or igneous cement, interstitial to cumulate pyroxene, olivine, or even plagioclase, in particular cumulate layers. For brevity, these deposits are referred to as concordant deposits in subsequent discussions.

Nelsonitic versus massive oxide facies. Nelsonite consists of ilmenite and apatite, commonly in proportions of about 2:1, and commonly is an equigranular medium-grained rock. Kolker (1982) found green spinel and zircon to be characteristic minor phases. Various varieties of nelsonite have been named on the basis of major impurities (Watson and Taber, 1913). As originally described by Watson and Taber, nelsonite occurred only as discordant bodies. However, concordant equivalents in ferrodioritic rocks have subsequently been discovered; several are described herein.

Nelsonitic facies deposits are the predominant type in some titanium-mineral districts, such as Roseland and San Gabriel Range. In other districts such as Allard Lake, Quebec, and Sanford Lake, they form subordinate facies where the major resources are massive-oxide facies.

Massive-oxide rock consists of coarse ilmenite with or without magnetite, both intergrown with other oxide-mineral phases.

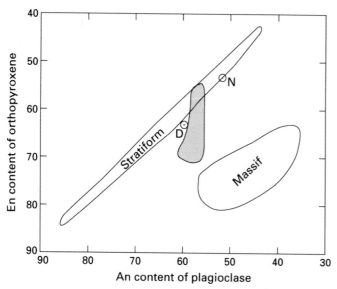

Figure 10. An-En diagram for stratiform versus massif anorthosites (from Anderson and Morin, 1969; see also Raedeke, 1982), showing compositions of Duluth Complex, Minnesota. Shaded field from South Kawishiwi anorthosites (M. P. Foose, written communication, 1986); N, Nathan's series; D, Duluth area, both averaged from Weiblen and Morey (1980, Fig. 5).

Other phases such as silicates and apatite may be minor; in fact, sulfides are locally the most abundant impurity. In some deposits, graphite is reported.

A transition between massifs and layered intrusions

Magmatic ilmenite deposits typically are associated with massif anorthosites and related rocks. In contrast, stratiform layered mafic intrusions do not have comparably valuable titanium-mineral deposits; much of the TiO_2 in these bodies is tied up in magnetite, and separable ilmenite is subordinate (see Chapter 5). Both magnetite and ilmenite are commonly cumulate in origin in layered intrusions.

Some igneous complexes containing anorthosites and ferrodiorites show characteristics that make them difficult to classify as either massif or stratiform type (cf., Romey, 1968). Two such bodies that contain titanium oxide deposits are described in the section on major deposits. Of these two, the rocks of San Gabriel Range are more closely related to the massif type, and the Duluth Complex is more closely related to the layered intrusions.

The anorthosite-syenite-ferrodiorite suite of the San Gabriel Complex is described by Carter (1982a; 1982b, p. 208) as both a layered complex and a massif. It is typical of an andesine anorthosite massif except for the orderly layering in the body, which consists of nine superposed layers. The body is also unusual in its lack of postintrusion metamorphism and deformation; this may have enhanced the preservation of layering.

The Duluth Complex (Weiblen and Morey, 1980) was em-

placed in an extensional environment in country rocks of low metamorphic grade. A nest of layered troctolitic intrusions in an older anorthosite series makes up the complex. Layering is well developed but lacks obvious order in terms of mineral composition, and no ultramafics are present. Late ferrodioritic intrusions are numerous. Some granulation of megacrysts in anorthosite is apparent. Mineral compositions are intermediate between those of massif and stratiform bodies (Fig. 10); plagioclase is about An_{55-60}, orthopyroxene about En_{55-65}, olivine about Fo_{55}. In some subunits of the complex, ilmenite deposits much like those of massif complexes are present.

Evidence of origin

If the magmatic ilmenite deposits had the compositions of ordinary rocks, the field evidence of intrusion (and their textures) would undoubtedly have been deemed sufficient to call them igneous rocks. For a long time, experimental evidence against liquids of this composition existing at reasonable temperatures prevented general acceptance of an igneous origin. However, Buddington and others (1955), Hargraves (1962), and Lister (1966) were not persuaded by this evidence and proposed not only igneous origins for these rocks but a parent liquid immiscible in the main silicate magma (another taboo for early experimental petrologists). More recent experimental evidence (Philpotts, 1967; Wiedner, 1982; Bolsover and Lindsley, 1983; Epler and others, 1986) shows that liquids of this composition may indeed be immiscible in ferrodioritic magma as it drops below 1,000°C, fluxed by either phosphate or elemental carbon. Thus, an igneous origin involving immiscible liquids for these rocks is respectable once more.

In more detail, the field relations and texture of the deposits support the immiscibility hypothesis and provide a glimpse of three stages in deposit formation. Figure 11 shows the relation among these stages in a cross section of the base of a crystalizing ferrodiorite sheet.

The first stage is the unmixing from cooling ferrodioritic magma of small spherical droplets of a titanium-rich liquid immiscible in the main parent magma. These droplets sink because of their great density. In most districts this first stage is texturally represented only by spherical inclusions in silicate phases (Fig. 12), probably as a result of the difficulties of preserving sinking droplets in mid-fall and of retaining droplet shape during crystallization of both oxide minerals and surrounding silicates.

The second stage is the arrival of these droplets at the magma floor, that is, at the top of the cumulate pile. This stage is recorded by the ilmenite deposits concordant to cumulate layering in ferrodiorite and related rocks. The deposits in San Gabriel Range of California provide an example.

These concordant ilmenite-rich bodies are present along the floors of ferrodiorite sheets and as cumulate-like layers within them (Fig. 11). Oxides (with or without apatite) are interstitial to cumulate phases and poikilitically enclose them (Fig. 13). This texture is present even in the thinnest cumulate layers (Fig. 14), as

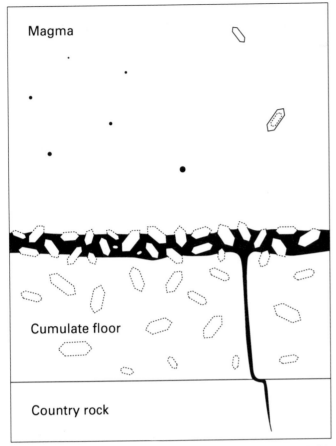

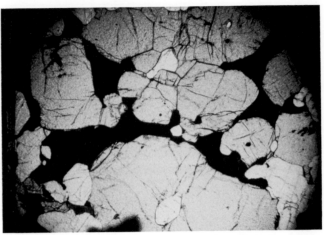

Figure 13. Photomicrograph of ilmenite, magnetite, and apatite enclosing cumulate olivine in layered ferrodiorite, Rattlesnake body, San Gabriel Range, California. Transmitted plane light, 6-mm field.

Figure 11. Diagram of a cumulate pile and overlying ferrodioritic magma chamber, showing the relation of three stages in the formation of magmatic ilmenite deposits. Solid pattern is dense immiscible liquid; crystals represent any cumulate solid, dotted where separated by other silicates.

first noted by Bateman (1951). Thus the concordant deposits of titanium oxides, though exceedingly cumulate-like in outcrop form, cannot be normal cumulates because texturally they are interstitial to known cumulate solids. This interstitial position in some described deposits is clearly not the result of metamorphism and deformation. Nor, given the geometry of deposition, can it be due to filter pressing or residual liquids. The evidence points rather strongly toward accumulation of a heavy liquid enriched in titanium, in the manner shown in Figure 11. Droplets, if supplied in sufficient quantity, coalesce to form an interstitial matrix in the uppermost part of the cumulate pile and poikilitically enclose cumulate phases. The titanium-rich interstitial fluid remains in the uppermost part of the pile, however, probably because poikilitic silicate crystallization has already filled interstitial positions farther down in the pile (Fig. 15). Therefore, this interstitial fluid remains concordant to cumulate layering in form. It is able to permeate even the narrowest selvages in uncemented cumulates, however, because of its extremely low viscosity (Kolker, 1982). Silicate liquid may be trapped (Fig. 16) in this interstitial network, just as cumulate solids may contain trapped droplets (Figs. 11, 12). Where droplet abundance is insufficient for titanium-rich liquids to coalesce in the interstitial spaces of the cumulate pile, silicate liquids may fill the remaining spaces and encase the droplets.

The presence in some districts of many discrete cumulate layers having interstitial oxide minerals, separated by normal cumulates, probably tells us that unmixing occurs only sporadically in the magma. Unmixing may be in response to additions of fresh magma to the chamber, a hypothesis that fits with the observed local coincidence of interstitial oxide minerals with cumulate olivine. In other layers in the same localities, however, the transitions from ordinary ferrodiorite to ilmenite-rich layers involve change only in the interstitial spaces; the cumulate phases remain the same (Fig. 15).

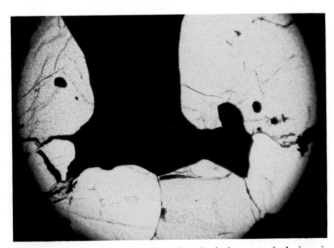

Figure 12. Photomicrograph of nearly spherical opaque inclusions in olivine, Rattlesnake body, San Gabriel Range, California. Transmitted plane light, 2-mm field.

Figure 14. Weathered cumulate ferrodiorite with thin ilmenite-apatite-cemented layers toward the top, Santa Clara Divide highway, San Gabriel Range.

The third and last stage in the formation of magmatic ilmenite deposits is the escape of this coalesced heavy liquid via fractures into lower structural units (most commonly anorthosite); here this liquid becomes a discordant magmatic ilmenite deposit (Fig. 11), such as those of the Allard Lake district of Quebec. This large deposit consists of virtually pure ilmenite-hematite rock in anorthosite. Locally, features indicative of mineral crystallization followed by resupply of fresh liquid are preserved (Fig. 17). Minor silicate impurities represent xenoliths and/or trapped liquids. These bodies are further described in a following section.

The sites of accumulation of titanium-rich liquids are probably structural depressions that grew on magma floors. Associated fractures aided liquid escape. In some districts, discordant magmatic ilmenite deposits form irregular intrusions that die out with depth (Bateman, 1951; Herz and Force, 1987).

The accumulation process, as outlined here, and the field evidence supporting it, are analogous to sulfide immiscibility in stratiform mafic bodies, such as that outlined by Scholtz (1936) and Naldrett (1979). Some deposits show textural evidence of both oxide and sulfide immiscibility.

CONTACT-METASOMATIC RUTILE DEPOSITS

Although magmatic ilmenite deposits commercially are the most important type of anorthosite-ferrodiorite titanium-mineral deposit, another type of deposit does exist. The two types are commonly assumed to be the same or closely related, but the second type is better regarded separately, as contact-metasomatic rutile deposits; they are analogous to skarn deposits. Magmatic ilmenite deposits originate with ferrodiorite and related rocks, whereas contact-metasomatic rutile deposits actually originate with anorthosite intrusion. The relations of anorthosite and ferrodiorite described in the preceding section imply that the contact-metasomatic deposits are everywhere older than magmatic ilmenite deposits found in the same complex.

The rutile albitite (kragerite) type of deposit is also described here, even though true anorthosite need not be present, because the mechanism of deposit formation is similar to such deposits in anorthosite. Kragerites have historically been likened to anorthositic rutile deposits (Watson, 1912; Green, 1956).

Rutile deposits on anorthosite margins

Alkalic andesine anorthosites contain rutile along their intrusive margins with older country rocks. Rutile is present in a marginal lithology that is coarse-grained, like anorthosite in the core of the body, but loaded with impurities such as pyroxenes and quartz. Rutile crystals are large in this zone; ilmenite may also be present. Rutile also is present in country rock within a few meters of the intrusive contacts; here rutile has approximately the

Figure 15. Photomicrograph of transition from layer having oxide-mineral oikocrysts to layer having silicate oikocrysts in ferrodiorite of Rattlesnake body, San Gabriel Range. Transmitted plane light, 6-mm field.

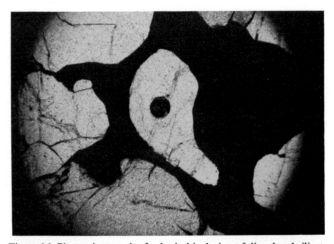

Figure 16. Photomicrograph of spherical inclusion of disordered silicate in apatite, representing trapped liquid, Rattlesnake body, San Gabriel Range. Transmitted plane light, 2-mm field.

Figure 17. Massive ilmenite-hematite ore of the Cliff ore body. Allard Lake district, Quebec. Exposure about 60 m high.

same distribution and grain size as ilmenite in unaltered country rock. Where swarms of anorthosite dikes and sills permeate country rock, deposits containing appreciable tonnages of rutile may form. Bodies of alkalic andesine anorthosite are presently known in only four places—Roseland and Montpelier, Virginia; Pluma Hidalgo, Oaxaca, Mexico; and St. Urbain, Quebec, Canada; rutile is present in all four deposits. Rutile deposits at Roseland and Pluma Hidalgo have the greatest economic potential.

Rutile mineralization is observed in these bodies along the upper margins of anorthosite bodies; hence the rutile is probably not cumulate in origin. The silicate megacrysts of the marginal facies locally lie athwart thin dikelets, making significant megacryst (and rutile) transport unlikely. The distribution of rutile in anorthosite and country rock seems to be most consistent with its formation in response to strong chemical gradients present during the intrusion of anorthosite at high temperatures (about 850°C or more for alkalic anorthosites) into country rocks that probably were recrystallizing under granulite-facies conditions. Many elements had sharp concentration gradients across the intrusive contact, but the most pertinent for rutile formation are iron and titanium. Anorthosite contains negligible amounts of both constituents. Titanium remained immobile, as it does in most geologic processes, but iron diffused into anorthosite along with other elements. In country rock, some ilmenite recrystallized to rutile. Adjacent to anorthosite, partial melting and recrystallization of country rock resulted in a coarse-grained rock containing the same mineral constituents as country rock, except coarse rutile that reflects diffusion of iron and only local redistribution of titanium. High metamorphic grade prevented this titanium from forming sphene.

Rutile-bearing albitites

Nearly monomineralic albite rocks, with rutile as the most important accessory, form small intrusives in two well-studied localities. At one locality, near Kragerø in southern Norway, this rock was named kragerite (kragerøite in modern Norwegian usage) by Watson (1912). One kragerite body was mined for rutile and is described separately. The other locality, in the Beaver Creek area of the northwestern Adirondack lowlands of New York, was noted by Brown (1983) and is first described here. It is of minor economic importance.

The two localities are strikingly similar in geologic setting, albitite relations and petrography, and controls of rutile formation. In both areas, early supracrustal rocks are isoclinally folded and show dome-and-basin configuration. Metamorphism is mostly of the upper amphibolite facies, with some metamorphic clinopyroxene. Both areas contain a variety of aplitic and pegmatitic intrusives, but rutile is limited primarily to albititic aplites. Host rocks of the albitites are biotitic amphibolites containing sphene and scapolite. Rutile albitites form mostly concordant but locally discordant intrusives up to 60 m thick; they commonly show gradational contacts with country rock by decrease of mafic minerals over widths of up to several meters. Xenoliths of country rock are extensively altered. Albitite is equigranular, consisting of 1- to 3-mm albite grains with minor quartz and variable microcline and tourmaline. Biotite is present near contacts with country rock. Typical rutile content is 1 to 2 percent; it occurs as equant grains and stubby prisms averaging 0.5 to 1 mm in diameter. The distribution of rutile around contacts of albitite with country rock reveals an important influence of metasomatism in these contact zones (Figs. 18, 19). This large number of features common to the two localities suggests the existence of a class of deposit, and implies quite an intricate control of the deposits by their geologic environment.

The New York occurrence may represent an ambient condition for this class of deposit, one well short of producing economically significant rutile deposits. Albitite bodies are as thick as 10 m, in zones of country rock 20 to 60 m thick (Fig. 19). Rutile content in albitite ranges from 1.0 to 2.3 weight percent in 15

Figure 18. Amphibolites and crosscutting albitites, Kragerø, Norway. The shape, size, and orientation of sphene clots in amphibolite are preserved as sphene-ilmenite-rutile clots in albitite.

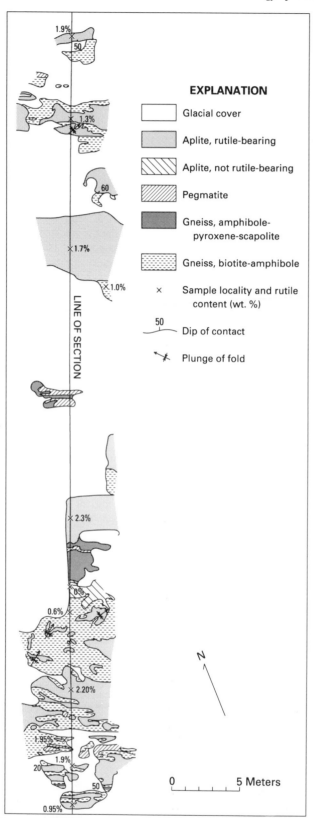

Figure 19. Strip map of one section through rutile albitite and its host rocks in the Beaver Creek area, New York. Bigelow Farm, near locality 50 of Brown (1983).

analyzed specimens and averages 1.7 percent. Rutile grain size averages 0.5 mm, with 75 percent coarser than 0.05 mm. Country rock within 1 m of albitite contacts averages 0.6 percent rutile, formed at the expense of sphene. In Figure 19, for example, 67 percent of the section consists of rutile-bearing rock, and average rutile grade for the entire section is 1.0 weight percent.

The distribution of rutile in deposits of this type shows that rutile forms at least partly by recrystallization of the titanium minerals in country rocks. Formation of ore-grade material, as at Kragerø, apparently requires that some horizons in country rock be extraordinarily high in TiO_2.

The formation of rutile in albitites shows both similarities to and differences from its formation in anorthosite-margin rutile deposits. In the case of albitite, there is an intrusive low in iron and calcium that destabilizes both ilmenite and sphene, under conditions where both are otherwise stable. In anorthosite-margin deposits, a low-iron intrusive destabilizes ilmenite, and granulite-facies conditions destabilize sphene.

MAJOR DEPOSITS

The following descriptions of anorthosite-ferrodiorite–related deposits include at least one important deposit of each subtype (Table 7). In addition, these descriptions constitute documentation of the geologic relations described previously.

Sanford Lake (Tahawus, MacIntyre) district, New York

The Sanford Lake district, now nearly inactive, is among the world's most important in terms of cumulative production. The district is in the heart of the Adirondack Mountains and on the southwestern edge of the Marcy massif of anorthosites and related rocks. The upstream end of the Hudson River per se is on one of the Upper Works, or Calamity Brook, oxide bodies.

Ilmenite has been mined in this district since 1942 by N L Industries and its predecessors. More than 10 million metric tons of ilmenite concentrate containing 46 percent TiO_2 have been produced. The remaining marginally economic resources are at least 20 million metric tons of ilmenite (Force and Lynd, 1984), divided among several deposits. The area mined most recently (until 1983) is the southern end of the Sanford ore body. The body having the largest remaining resources is near Cheney Pond, 2 km west of the Sanford body. Substantial resources may also be present in the Upper Works–Calamity Brook area 5 km to the north.

Anorthosite is the predominant rock of the district, ranging in composition from An_{43} to An_{55} and averaging An_{50} (Ashwal, 1982a). Antiperthitic lamellae are common. Two facies are present: the Marcy Anorthosite containing andesine and labradorite megacrysts, and the Whiteface Anorthosite, which is finer grained, more sodic, more mafic, and foliated. The Whiteface is a marginal facies along the outside of the domical massif and structurally overlies the Marcy facies.

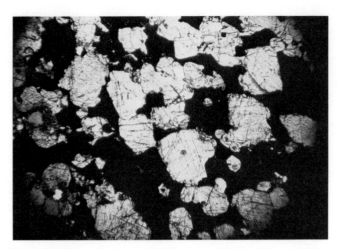

Figure 20. Photomicrograph of oxide minerals interstitial to cumulate orthopyroxene in gabbro-hosted layers of an Upper Works body, Sanford Lake district, New York. Transmitted plane light, 6-mm field.

Gabbro[2] is subordinate to anorthosite in areal extent in the Sanford Lake district, but gabbro is spatially associated with ilmenite deposits. There is a continuous gradation from gabbro to ilmenite ore by increase in the number and thickness of oxide-rich layers concordant to fine layering in gabbro. Gabbro intrudes anorthosite and contains xenoliths of it (Stephenson, 1945; Gross, 1968).

In the Sanford Lake–Sanford Hill belt of deposits, which is about 1.5 km long, gabbro structurally overlies anorthosite (Gross, 1968). Ilmenite deposits are associated with both rock types in such a way that gabbro-hosted deposits are called hanging-wall deposits at the mine. The economic gabbro-hosted deposits are concordant to layering in gabbro and are localized along the base of gabbro bodies. Thicknesses of such deposits are as great as tens of meters (Gross, 1968), but some ilmenite-enriched layers are only a few grains thick. Ilmenite in these deposits is typically 1 to 2 mm in diameter and is more abundant than magnetite (Stephenson, 1945; Gross, 1968). The host gabbros contain from less than 1 to 10 percent apatite and much disseminated metamorphic garnet. Ashwal (1982a; see also Bateman, 1951) noted that the typical texture of gabbro rich in oxide minerals has cumulate pyroxene with interstitial oxides (cf., Fig. 20); individual cumulate layers grade upward from oxide rich to oxide poor. Within oxide-mineral domains, magnetite and ilmenite grains show an interlocking texture (Fig. 21). Pyroxene, garnet, apatite, and other gangue minerals in gabbro-hosted ilmenite ore together average about 30 wt %.

Ilmenite ore in anorthosite of this belt of deposits is highly discordant, forming dikes and sills of all thicknesses, with apophyses (Stephenson, 1945; Gross, 1968). Xenoliths of anortho-

site in this type of ore are common; indeed, I have seen intrusion breccias of anorthosite blocks veined by ore. At the scale of geologic maps, however, most of the larger lenses of ore in anorthosite are parallel to ore lenses in adjacent gabbro (Gross, 1968). Garnet selvages are present between ore and anorthosite host. Ilmenite grain size is typically 2 to 3 mm. Ilmenite is less abundant than magnetite in anorthosite-hosted deposits (Stephenson, 1945; Gross, 1968). Ore of this type is typically almost free of silicate gangue. Ore bodies may be up to 60 m thick (Stephenson, 1945). To my knowledge, economic bodies are nowhere more than 100 m structurally beneath gabbro in this belt.

Ilmenite-rich bodies of considerable size are also present in the Upper Works–Calamity Brook area. They form tabular bodies concordant to gabbro and discordant to anorthosite, similar to those of the Sanford deposits (Stephenson, 1945). Their resources were not evaluated by Gross (1968) and have not been reported. Gabbro-hosted ilmenite enrichments show textures as in the Sanford deposits; oxide minerals are interstitial to cumulate phases (Fig. 20).

Ilmenite deposits in the Cheney Pond area are hosted entirely by gabbro that is intrusive into anorthosite (Stephenson, 1945). Gross (1968) reported that gently dipping gabbro is both underlain and overlain by anorthosite. The thickest portion of the gabbro that is rich enough in ilmenite to constitute ore occurs in a minor syncline. The gabbro is finely banded; layers dominated by oxide minerals constitute about 20 percent of the rock sequence and may be more than 1 m thick. The nelsonite collected by Kolker (1982) near Cheney Pond is now shown by extensive test-pitting to be a minor lithology.

In all these deposits, the oxide minerals consist of ilmenite with variable finely exsolved hematite and of magnetite with considerable amounts of finely exsolved ilmenite and spinel (Ashwal, 1982a). The magnetite contains up to 3 percent V_2O_5 (Balsley, 1943; Ashwal, 1982a).

Figure 21. Photomicrograph of ilmenite-magnetite (pitted) relation in gabbro-hosted ore of the south end of the Sanford ore body. Reflected light, 3.5-mm field.

---

[2]As described in the literature; actually ferrodiorite (P. Ollila, written communication, 1986).

Allard Lake (Lac Tio) district, Quebec

The Lac Tio deposit of the Allard Lake district, discovered in 1946, has produced more than 40 million metric tons of ore, and larger quantities remain in the ground. At present, this single deposit supplies 19 percent of the world's titanium needs (Table 4). An innovative ore-smelting technique has given the ilmenite-hematite deposits of this district a competitive edge. The Lac Tio deposit is inland 40 km from Havre St. Pierre, on the northern shore of the St. Lawrence seaway (Fig. 1). The mine site itself is an uninhabited, heavily glaciated, lightly forested area of moderate relief, reached via company railroad.

The Allard Lake district, described by Hargraves (1962) and Bergeron (1972, 1986) and mapped by Hocq (1982), contains about six important titanium-mineral prospects. The district is toward the eastern edge of a large andesine anorthosite massif, with several inliers of later ferrodiorite rich in apatite and oxide minerals. These ferrodiorite bodies contain massive ilmenite en-

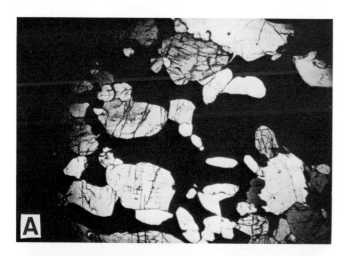

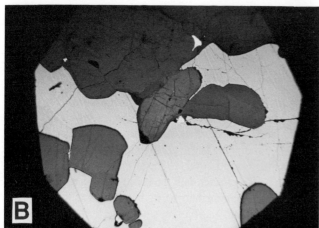

Figure 22. Photomicrographs of ilmenite enrichments in ferrodiorites of the Allard Lake district, Quebec. A. Interstitial texture of oxide minerals and apatite. Cumulate phases in the field of view are orthopyroxene and plagioclase. Transmitted plane light, 8mm field. B. Same texture in reflected light, showing exsolved hemoilmenite poikilitically enclosing cumulate minerals, 3.5-mm field.

richments along their bases. Ferrodiorite consists mainly of plagioclase (An$_{41-44}$, a composition related by Hargraves and Bergeron to contamination from anorthosite), orthopyroxene, 8 to 10 percent apatite, and ilmenite and lesser magnetite together totaling at least 20 percent by weight. Coarsely intergrown ilmenite and hematite, along with magnetite, apatite, and sulfides, occupy interstices (Fig. 22) between cumulate phases and poikilitically enclose them.

Hargraves (1962) and Bergeron (1986) relate massive ilmenite-hematite ore to ferrodiorite parents. All observers seem agreed that the massive oxide rock is igneous in origin; indeed, Hargraves (1962) and Lister (1966) founded their pioneering hypothesis of liquid immiscibility for magmatic ilmenite deposits on evidence from the Allard Lake district. This is historically remarkable because the evidence is so much better elsewhere.

The Lac Tio deposit has been described by Hammond (1952). It is a massive, coarse-grained, subhorizontal igneous sheet more than 60 m thick, in anorthosite. It consists essentially of platy crystals measuring $10 \times 10 \times 2$ mm composed of coarse intergrowths of ilmenite (75 percent) and hematite (25 percent). This ore contains 32 to 36 percent TiO$_2$. Minor constituents of ore include magnetite, sulfides, apatite, hercynitic spinel, and zircon. Magnetite is less abundant than sulfides, which here include several Ni-Co minerals. Rose (1969) presented a K-Ar age of 1025 Ma for a pegmatite that cuts anorthosite but is cut by ore of the Lac Tio deposit.

A part of the orebody known as the Cliff dips gently east and contains silicate-bearing bands (Hammond, 1952). The form of these bands (Fig. 17) is analogous to a fluvial channel deposit with festoon cross beds. In this case, movement and precipitation of oxide magma in a passageway through anorthosite produced the structures.

Geometric features of the deposit are consistent with derivation of oxide magma from nelsonitic components immiscible in ferrodioritic magma. The presence of apatite and magnetite in concordant enrichments in ferrodiorite (Fig. 22) but not in discordant massive ore is unexplained.

Tellnes district, Norway

The Tellnes and Storgangen deposits of the southern coast of Norway have been important titanium-mineral producers since 1902. The Storgangen deposit closed in 1964 and has been supplanted by the Tellnes deposit, discovered in 1954 about 2 km to the south. At present, this deposit supplies 12 percent of the world's titanium. Total production from the two deposits has been more than 19.5 million metric tons of ilmenite, and remaining reserves are more than 140 million tons (Krause and others, 1986; Korneliussen and others, 1986). Ilmenite concentrates contain 44 to 45 percent TiO$_2$.

The district is in jumbled coastal mountains of the Åna-Sira anorthosite massif (Fig. 23). It consists of three main deposits: Tellnes, Storgangen, and Blåfjell; the geology of all three has been summarized by Krause and others (1986). Geologic maps of the

Figure 23. View from the western end of Tellnes mine, Norway, with mountains of Åna-Sira anorthosite massif in background. Anorthosite (A) overlies dark-colored ore (O) forming a J-shaped cylindrical intrusion in this down-plunge view (opposite to view in Fig. 24).

district are by Krause and others and by Falkum (1982). The Åna-Sira massif consists of andesine anorthosites and interlayered leuconorites that are dated at about 900 Ma (Pasteels and others, 1979), are intrusive into high-grade metamorphic rocks (Wilmart and Duchesne, 1987), and are structurally overlain by the Bjerkreim-Sokndal lopolith (Duchesne, 1972). This layered intrusion consists of lower leuconorites and of upper monzonorites with unusually abundant ilmenite and magnetite (in interstitial positions), fayalite, and apatite (Duchesne and others, 1987). The monzonoritic rocks should be called ferrodiorite, based on their mineralogy, the composition (antiperthitic, An <40) and abundance (>50 percent) of feldspar, and their high Fe/Mg ratios (from Duchesne, 1972, Fig. 2; Krause and others, 1986, Table 2).

Tellnes deposit. This deposit is a steeply dipping discordant tabular intrusion of ilmenite norite[3] in anorthosite. Contacts with anorthosite are sharp, and intrusion breccias and anorthosite xenoliths are locally present along it. The orebody is 2,700 m long and as much as 400 m wide. In cross section it is a J-shaped body (Figs. 23, 24) in anorthosite, generally dipping steeply to the south (Ragnar Hagen, written communication, 1987); the true thickness of ore is 200 to 250 m. This shape is apparently an original feature of intrusion; it plunges gently east-southeast under anorthosite. Ore is associated and apparently coeval with a set of mangeritic dikes that physically link the Tellnes body with the lopolith. Mangeritic rock also forms a marginal zone of the ore intrusion.

The ore averages 18.4 percent TiO_2 and contains (by volume) 53 percent homogeneous andesine averaging $An_{44}Or_{3.6}$ in composition, 29 percent (39 percent by weight) ilmenite with hematite and spinel lamellae, and 10 percent orthopyroxene, with

lesser clinopyroxene, olivine, and biotite (Table 8). Magnetite, apatite, sulfide minerals (including Ni-Co sulfides), and baddeleyite are minor constituents; magnetite is localized in one zone (Fig. 24). The baddeleyite is included within ilmenite (Gierth and Krause, 1974).

Tellnes ore appears quite homogeneous and typically has a subophitic texture formed by orthopyroxene and plagioclase grains up to about 2 mm long. Clinopyroxene is interstitial, and brown amphibole forms small oikocrysts apparently nucleated by olivine. In drill core, alternating zones of oxide-rich and oxide-poor rock are visible (Ragnar Hagen, written communication, 1987).

Enclosed by this subophitic texture of silicate minerals is another texture that contains most of the ilmenite in the rock (Fig. 25). In this texture, globules about 0.2 to 0.5 mm in diameter consist mostly of single ilmenite crystals and their hematite exsolution lamellae, with subordinate magnetite, sulfides, apatite, green spinel, and baddeleyite (Table 8). Some globule shapes are slightly modified by ilmenite crystal faces. Two or more globules are commonly in contact, thus forming a texture analogous to bunches of grapes[4] (Fig. 25A); adjoining globules have planar mutual walls. The "grape-bunch" texture may be enclosed either in orthopyroxene or by the outer portions of plagioclase crystals, and triaxial interglobule spaces are also filled by orthopyroxene or plagioclase (Fig. 25). The cores of plagioclase crystals are free of globules, except in marginal zones of the orebody. In some specimens, magnetite is present both in globules and as coarser separate anhedral grains. In specimens particularly rich in oxides, both ilmenite and magnetite are coarse and anhedral, and the "grape-bunch" texture is not apparent. Minor late ilmenite shows other morphologies (Gierth and Krause, 1973).

The marginal zone of ilmenite norite, in contact with anorthosite, shows a variation of globule morphology that probably represents morphology at intrusion, encased in the earliest crystallized silicates (Fig. 25B). Antiperthitic plagioclase nucleated on the contact encloses a variety of spherical, ellipsoidal, and budding globules in a wide range of sizes.

The oxide-rich globular domains may represent former droplets of immiscible oxide-rich liquid. If so, high surface tensions and high silicate magma viscosity must have retarded droplet merger until silicate crystallization preserved them.

Storgangen deposit. The Storgangen deposit is also an elongate discordant intrusion in anorthosite, subparallel in strike to Tellnes but dipping to the north. It has been described in some detail by Krause and Pape (1975, 1977). At the western end of the body, it is in contact with the base of the Bjerkreim-Sokndal lopolith and is concordant to it, but eastward the body cuts down-structure into anorthosite, which thus forms both the hanging wall and footwall of the body. The Storgangen body is offset by mangerite dikes that are related to the Tellnes body.

---

[3]Again, I would call this rock ilmenite ferrodiorite based on its mineralogy and chemistry as reported by Krause and others (1986). Henceforth I will simply call it *ore.*

[4]This texture is not, however, the same as botryoidal texture in authigenic minerals.

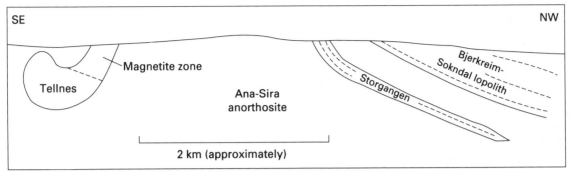

Figure 24. Schematic relation of Tellnes and Storgangen deposits and base of Bjerkreim-Sokndal lopolith in NW-SE cross section. Restoration of the latter two to nearly horizontal rotates the Tellnes body to nearly vertical, with a nearly horizontal mineralogic boundary. Internal layering dashed.

TABLE 8. BULK CHEMISTRY, MODAL ANALYSES, AND MINERAL CHEMISTRY (WHERE KNOWN) OF TELLNES AND STORGANGEN BODIES, NORWAY*

Modes (vol. %)	Storgangen	Tellnes
Quartz	0	0
Plagioclase	52.3[†] (An 43–55)	53.2 (An 35–55)
Orthopyroxene	? (Fs 25–30)	10.2 (Fs 20)
Clinopyroxene	P	0.8
Biotite	P	3.9
Olivine	...	0.9
Apatite	P	0.5
Ilmenite	22.2[†]	28.6 (6.1% MgO)
Magnetite	P	0.7 (0.7% V_2O_3, 0.7% Cr_2O_3)
Sulfide	P	0.5
Baddeleyite	P	P

Chemistry (wt. %)	Storgangen[§]	Tellnes
SiO_2	27.9	30.4
TiO_2	19.6	18.4
Al_2O_3	8.3	11.7
Fe_2O_3	11.7	7.3
FeO	19.2	18.1
MgO	7.0	6.1
CaO	3.3	4.4
Na_2O	n.a.	2.4
K_2O	n.a.	0.6
S	0.3	0.2
P_2O_5	0.04	0.09

*From Krause and others (1986), Krause and Pape (1975, 1977), and Gierth and Krause (1973, 1974). P = present; n.a. = not analyzed.
[†]Average of 40 specimens from throughout the section, calculated from Krause and Pape (1977, Fig. 2).
[§]Integrated composition from 1963 average mill feed; Ragnar Hagen, written communication, 1988.

In contrast to the Tellnes body, the Storgangen deposit is strongly layered (Fig. 26) throughout its 50-m thickness. Oxide minerals are more abundant in dark layers near the base of the body. Bulk composition is similar to the Tellnes body (Table 8), but layers vary from leuco-"norite" to ilmenite "norite."

Oxide minerals are interstitial to cumulate crystals of orthopyroxene and plagioclase throughout the Storgangen deposit (Fig. 27; see also Krause and Pape, 1977, Fig. 15). Ilmenite only locally encloses pyroxene poikilitically here, as interstitial domains alternately consist of ilmenite, magnetite, and minor green spinel. Minor apatite and sulfides are also present (Table 8). The texture of this body suggests that dense immiscible liquid accumulated in the cumulate pile.

Relation of the Tellnes and Storgangen deposits. Table 9 compares the compositions of the interstitial domains at Storgangen with the globular domains at Tellnes. These compositions are sufficiently close that the Tellnes globules could plausibly represent an early stage of evolution of an immiscible liquid similar to that represented at Storgangen.

Cross-cutting relations suggest that Storgangen is slightly older than Tellnes, but both are closely linked to the Bjerkreim-Sokndal lopolith, and their compositions are similar. The plane defined by the Storgangen body is nearly parallel to the base of the gravity-layered lopolith (Fig. 24), so Storgangen should also have differentiated by gravity. If the plane of the Storgangen body is restored to an approximately horizontal attitude, most of the Tellnes body becomes nearly vertical (Fig. 24). Thus the Tellnes body might be expected to show little cumulate layering. Probably Tellnes cooled and crystallized mostly inward from nearly vertical walls. If immiscible heavy liquid droplets were present, they would be intercepted in their fall and enclosed in plagioclase-seeded ophitic intergrowths crystallizing inward from these walls.

Bjåfjell deposits. The ilmenite ores at Blåfjell were worked for iron before either the Storgangen or the Tellnes deposit. The Blåfjell deposits, described by Krause and Zeino-Mahmalat (1970) and Krause and others (1986), are associated with a coarse ferrodiorite ("norite-pegmatite") intruded into anorthosite

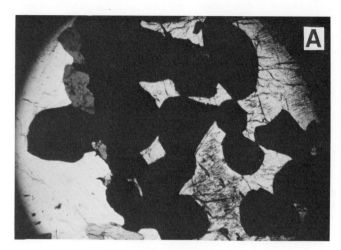

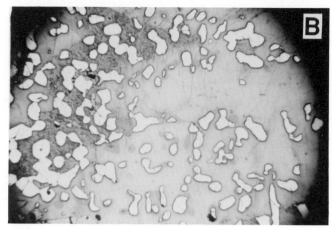

Figure 25. Photomicrographs of globular hemoilmenite crystals (and other minor phases) enclosed by subophitic orthopyroxene and plagioclase, Tellnes ilmenite norite. A. "Grape-bunch" texture of hemoilmenite in orthopyroxene. Transmitted plane light, 2-mm field. In reflected light, globules have planar mutual walls that bound different crystal directions in hemoilmenite. B. Globules in rapidly cooled zone within 1 cm of anorthosite contact. Reflected light, 6-mm field. Darker host is altered pyroxene.

Figure 26. Layering in Storgangen ore. Plagioclase and orthopyroxene in different proportions are the main cumulate phases.

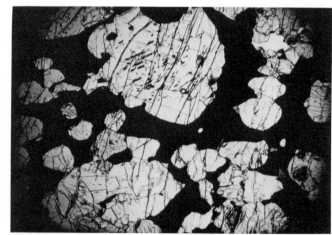

Figure 27. Photomicrograph of interstitial texture of oxide minerals relative to cumulate orthopyroxene, Storgangen. Transmitted plane light, 6-mm field.

TABLE 9. PERCENTAGES OF MINERALS IN INTERSTITIAL DOMAINS AT STORGANGEN AND IN GLOBULAR DOMAINS AT TELLNES, BASED ON 1,000-POINT COUNTS.

	Base of Storgangen body		Contact zone of Tellnes body	
	% rock (vol.)	% domain (vol.)	% rock (vol.)	% domain (vol.)
Silicate minerals	72.2		69.0	
Domain	27.8		31.0	
Ilmenite-hematite		84.4		86.3
Magnetite		14.0		6.1
Green spinel		0.9		0.3
Sulfide minerals		0.6		1.3
Apatite		0.1		3.2

along the crest of the Åna-Sira dome. The coarse-grained ilmenite-hematite ores are, in my opinion, of two types: (1) layered ores along the ferrodiorite-anorthosite contact (originally the base of ferrodiorite), with coarse ilmenite interstitial to cumulate coarse euhedral plagioclase (Fig. 28); and (2) discordant, nearly monomineralic, ilmenite bodies in anorthosite, apparently with strongly linear shape in the Undergruve mine workings. The relation of the Blåfjell deposit to the Tellnes and Storgangen deposits is not clear.

Roseland district, Virginia

The Roseland district contains examples of both contact-metasomatic rutile deposits and magmatic ilmenite deposits. The ilmenite deposits are of the nelsonitic type and contain both concordant (ferrodioritic host) and discordant (anorthositic host) deposits. Deposits of all of these types have been mined, but no deposits are being mined currently. Remaining resources are approximately 1.5 million metric tons of rutile and 12.5 million metric tons of ilmenite (Herz and Force, 1987). The ilmenite of the district is nearly stoichiometric in composition and commonly is free of intergrowths. The following descriptions are after Herz and Force; earlier descriptions of the district are by Watson and Taber (1913) and Ross (1941).

The district consists of a domical alkalic andesine anorthosite intruded along its upper and outer margins into metasedimentary and metavolcanic gneisses. Anorthosite and gneiss share a fabric showing granulite-facies metamorphic assemblages. These units are structurally overlain by younger ferrodiorites that are unusually rich in silica and potash. The age of anorthosite is about 1,050 m.y.; the ferrodiorites are about 980 m.y. old.

Coarse-grained rutile is developed along the contacts of anorthosite with older country rock, especially the ilmenite-bearing metavolcanic gneiss. Rutile is present in a marginal impure facies of coarse anorthosite and in country rock, in a zone that straddles the contact. Where there are swarms of anorthosite

dikes and sills, rutile is almost regional in extent. Rutile concentrations typically are only about 2 percent by weight, which approximately matches the TiO_2 content of metavolcanic country rock. In some small deposits, however, rutile exceeds 2 percent; at one of these deposits, the Roseland Rutile mine, rutile was mined until 1949.

Both rutile and ilmenite are present in these deposits; ilmenite increases at the expense of rutile passing into country rock. Ilmenite is present as single crystals without appreciable intergrowths.

The temperature of intrusion is thought to have been about 850 °C based on two-pyroxene geothermometry. Country rock must have been at granulite-facies temperatures, here also of about 850 °C, either during intrusion or shortly afterward.

The two ferrodiorite units are younger and form igneous sheets concordant to the regional domal structure. However, numerous ferrodiorite dikes are present in the older, structurally underlying units. The ferrodiorites contain xenoliths of anorthosite and granulite gneiss. The ferrodiorites contain one fewer metamorphic fabric than the older anorthosite and granulite gneiss and show only amphibolite-facies recrystallization. The ferrodiorites characteristically have high apatite and ilmenite contents.

Concordant impure nelsonites form discontinuous bodies along the bases of the ferrodiorite sheets. These bodies consist of tabular cumulate layers with ilmenite-apatite net veins poikilitically enclosing cumulate pyroxenes (Fig. 29). Also present in the bodies are elliptical leucocratic quartz-feldspar-pyroxene domains flattened in a direction parallel to the base of ferrodiorite sheets. Large ilmenite-rich bodies of this type along the bases of ferrodiorite sheets were mined until 1971. The texture of the mined bodies is not known, however, because they are intensely weathered.

Figure 28. Hand specimen of Blåfjell ore (Norway) showing coarse oxide minerals interstitial to euhedral cumulate plagioclase.

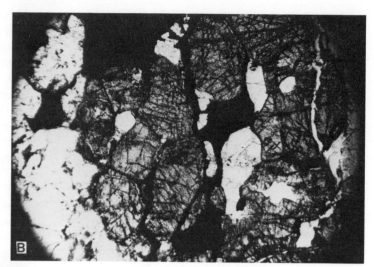

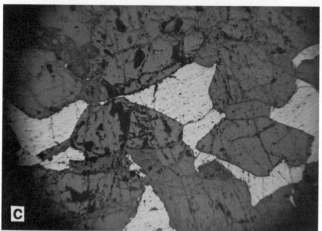

Figure 29. Elliptical domains of felsic minerals among cumulate ortho-pyroxene and interstitial ilmenite-apatite, Roseland district, Virginia. A. Hand specimen; field of view is 50 mm. B. Photomicrograph of mafic domain showing net-vein texture of interstitial oxides and apatite among cumulate orthopyroxene grains. Transmitted plane light, 1.8-mm field. C. Same texture in reflected light, showing poikilitic enclosure of cumulate pyroxene by hemoilmenite, 1.8-mm field.

The nelsonite bodies for which the rock type was named occur as discordant intrusions. Most of these are too irregular in shape to properly be called dikes. Most of them are small; I have seen discordant nelsonites thinner than 1 cm. Many bodies consist of only ilmenite and apatite, as equigranular medium-grained rock. There are, however, all transitions from pure nelsonite to ilmenite- and apatite-rich ferrodiorite intrusions. Some of the latter have thin cumulate nelsonite accumulations along their bases.

All of the discordant nelsonite bodies occur in country rock just below the base of ferrodiorite intrusive sheets. Several are known from drilling or from stream-gorge outcrops to die out at depth. Only the weathered overburden over the thickest nelsonite body has been extensively mined, near the town of Piney River. This body has abundant chlorite and minor sulfide impurities. Significant resources remain in fresh rock of this body.

San Gabriel Range, California

The anorthosite-syenite-ferrodiorite suite of the San Gabriel Complex, described and mapped by Higgs (1954), Oakeshott

(1958) and Carter (1982a, b), contains numerous ilmenite pros-pects of two types: Discordant ilmenite pyroxenites are present in anorthosite, and concordant bodies enriched in ilmenite, magne-tite, and apatite are present in ferrodiorite. Economic interest in the district has recently been rekindled after a virtual lapse of more than forty years (Industrial Minerals, 1986).

The district forms much of the crest of the western portion of the San Gabriel Range, immediately north of the Los Angeles suburbs (Fig. 30), and is largely in national forest and shooting preserve. Slopes are steep and the climate semiarid. The range is caught between the San Andreas and San Gabriel faults, so frac-turing is locally severe. In coherent domains between fractures, however, original textural relations are unusually well preserved, because metamorphism and plastic deformation of the complex are minimal. It is for this reason that textures from the San Gabriel Range were used as an example earlier in this chapter.

The igneous complex dates from about 1200 Ma and covers about 250 km^2. It consists of a basal anorthosite unit, a syenite unit, and an upper ferrodiorite unit (jotunite of Carter, 1982a). The complex seems intermediate in character between massif and stratiform types. Country rock of the complex is the Mendenhall

Figure 30. The San Gabriel Range from Santa Clara Divide, looking south into the San Fernando Valley, California. Ferrodiorite occupies the foreground and Pacoima Creek (first valley on left side). Mendenhall Gneiss occupies the next ridge and San Gabriel Fault, the next valley.

Gneiss, a finely banded gneiss of granulite metamorphic facies. The igneous complex is arranged in a rude dome, and the upper ferrodiorite unit is in exposed intrusive contact with the gneiss.

The anorthosite unit, the oldest of the complex, consists of coarse calcic andesine. Leucogabbro is locally present in the anorthositic unit and retains a subophitic texture. Overlying syenite is similar to anorthosite but has mesoperthitic feldspars. The ferrodiorite unit, the youngest, characteristically contains plagioclase, commonly antiperthitic; two pyroxenes; ilmenite; magnetite; and apatite. Several of its five subunits (of Carter, 1982a) show cumulate layering, defined by grain size and/or mineralogic transitions, commonly graded.

Ilmenite deposits in anorthosite and Mendenhall Gneiss are present as small discordant ilmenite pyroxenites. Apatite contents of these deposits vary from 2 to 20 percent.

The concordant ilmenite deposits in ferrodiorite are in cumulate layered rocks. My experience with these deposits is largely with the so-called Rattlesnake and Saturday Night claims in Pacoima Creek and the bodies exposed along the Santa Clara Divide highway. High-grade layers range in thickness from one cm (Fig. 14) to more than 10 m. Ilmenite forms more than 20 percent by weight of some layers, and can be concentrated into fractions containing 45 percent TiO_2. The bodies share one important textural characteristic (Force and Carter, 1986): ilmenite, magnetite, and apatite form net vein systems interstitial to cumulate phases in particular layers (Fig. 13). Green spinel may also form part of the interstitial assemblage and commonly separates large single crystals of magnetite and ilmenite. Ilmenite poikilitically encloses cumulate phases and is typically free of intergrown oxides. The edges of the bodies are defined by the substitution of an interstitial silicate phase, poikilitically enclosing cumulate phases, for ilmenite, magnetite, and apatite in the same positions (Fig. 15).

The cumulate phases are commonly rounded euhedra and may show textural grading within an ilmenite-rich body. Olivine is the cumulate phase in many of the ilmenite-rich layers; olivine is not common in the ferrodiorite elsewhere. The cumulate phases contain spherical ilmenite inclusions; locally these inclusions are arranged in rings that apparently outline growth stages of the cumulate grain. Conversely, apatite shows spherical inclusions of disordered silicate (Fig. 16).

Laramie Range, Wyoming

An anorthosite massif about 1,400 m.y. old (Smithson and Hodge, 1972) forms much of the Laramie Range. Plagioclase composition ranges from An_{40} to An_{65} and averages about An_{55}. The body has been described and mapped by Newhouse and Hagner (1957) and Klugman (1966). Little ferrodiorite or gabbro has been reported, but numerous rocks enriched in iron-titanium oxide minerals form sharply discordant to concordant lenses in labradorite anorthosite (Diemer, 1941; Goldberg, 1984). Enriched rocks vary from oxide-rich anorthosite, leuconorite, and troctolite to younger olivine-bearing massive magnetite-ilmenite rock. Apatite is present in some enriched bodies (Bolsover and Lindsley, 1983).

The largest of the enriched bodies, at Iron Mountain, was mined for heavy aggregate in the 1960s. This deposit has been described by Pinnell and Marsh (1954), Hagner (1968), and Eberle and Atkinson (1983) and mapped by Newhouse and Hagner (1951) and Dow (1961). Massive magnetite-ilmenite rock crosscuts an east-dipping contact zone between structurally overlying troctolite and underlying leucogabbro. The troctolite is locally rich in interstitial magnetite and ilmenite and contains apatite. Massive magnetite-ilmenite rock contains as much as 50 percent olivine. Ilmenite is present both as intergrowths in magnetite and as coarse separate grains containing exsolved spinel (G. Turner in Hagner, 1968). The olivine-poor rock toward the core of these bodies has a high ratio of magnetite to ilmenite, and its V_2O_5 content averages 0.64 percent (Dow, 1961).

Massive magnetite-ilmenite rock from Iron Mountain contains about 20 percent TiO_2, but acceptable ilmenite concentrates have not yet been made from it (Pinnell and Marsh, 1954; Dow, 1961). Thus, it is questionable whether the deposit should be listed as a resource using the Force and Lynd (1984) definition.

Duluth Complex, Minnesota

The Duluth Complex consists of a series of anorthositic, gabbroic, and troctolitic intrusions dating from about 1100 Ma (Weiblen and Morey, 1980). Titanium-mineral resources of the Duluth Complex are of two types. The first is ilmenite recoverable as a byproduct of mining for base-metal sulfides in troctolitic intrusions (Iwasaki and others, 1982). The ratio of ilmenite to magnetite varies among the sulfide deposits (Pasteris, 1985). Sulfide and some oxide minerals are present in interstitial positions in these rocks (Fig. 31; see also Pasteris, 1985, Fig. 6).

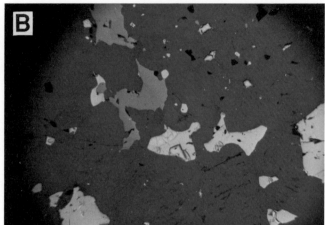

Figure 31. Photomicrographs of textural relations of oxide and sulfide minerals in rocks of the Duluth Complex, Minnesota. A. Interstitial sulfide and oxide minerals (both opaque) at the base of troctolitic layered intrusion, Dunka Pit. Cumulate crystals are mostly clinopyroxene (bottom in this view) and olivine (top). Transmitted plane light, 3.5-mm field. B. Same view in reflected light. Ilmenite gray, sulfides bright. C. Interstitial pyroxene with cumulate plagioclase, olivine, and ilmenite in unit G of Nathan (1969). Transmitted plane light, 3.5-mm field.

The other type of resource, of economic interest for titanium alone, is mostly confined to one small part of the Duluth Complex in the layered series of Nathan (1969), adjacent to the Canadian border. A separate intrusion, the Water Hen intrusion, is mentioned by Nafziger and Elger (1987) as containing additional resources of this type.

The layered series of Nathan (1969), probably the oldest of the complex (Weiblen and Morey, 1980), is systematically enriched in iron-titanium oxides. It is in a heavily glaciated, remote area of low relief with a maze of lakes. The series consists of successive intrusions of troctolitic and gabbroic compositions, with minor quartz and K-feldspar in a few lithologies. The earlier intrusions are concordant sheets dipping gently south, and a later set of intrusions is discordant. Nathan's unit G of coarse gabbro is the most extensive unit rich in oxide minerals and is among the youngest concordant bodies. Nathan's discordant units T and V are closely associated with unit G and contain even higher contents of oxide minerals.

Unit G, which is as much as 400 m thick, shows fine modal-graded layering. Major cumulate phases are plagioclase, olivine, and augite. Plagioclase composition in unit G trends from An_{54} at the base to An_{50} at the top. No other simple trends in mineral composition were found by Nathan (1969).

The oxide minerals of unit G vary from about 10 to 60 percent by volume; apatite reaches 5 percent in the uppermost of three subunits. Nathan (1969) considered the oxide minerals cumulate, whereas Grout (1949–1950) considered them interstitial and therefore postcumulate. Lister (1966) proposed that the oxide minerals represent a heavy immiscible liquid that crystal-lized in interstices. My specimens suggest that all parties are partly correct; some rocks are plagioclase-olivine-ilmenite cumulates with interstitial pyroxene and lesser oxide and sulfide minerals (Fig. 31).

The oxide minerals of unit G consist of ilmenite and magnetite in coarse grains; ilmenite is largely free of included lamellae but may contain minor magnetite, ulvospinel, and hercynite (Nathan, 1969). Magnetite contains abundant lamellae of ilmenite and ulvospinel. Lister (1966) found that this magnetite contains 0.3 to 0.4 percent vanadium. The ratio of magnetite to ilmenite averages about 1:1 but varies with stratigraphic position within the unit. Lister (1966, p. 293) found magnetite predominant over ilmenite in three out of the four oxide-rich layers of unit G.

Units T and V are small discordant bodies containing abundant olivine and augite, respectively. Coarse oxide minerals vary from 20 to 70 percent by volume. Apatite contents are minor. Separable ilmenite is only up to 10 percent; that is, the ratio of magnetite to ilmenite is high in the rocks richest in oxide minerals. Intergrowths in the oxide minerals are similar to those described for unit G.

Grout (1949–1950) appraised the economic potential of titanium minerals in this area. He focused on oxide mineral en-

Figure 32. Rutile-rich rock of the Agua Titania deposit, Pluma Hidalgo district, Mexico, in hand specimen view with rutile reflectant. Scale in millimeters.

richments subsequently mapped as part of unit G and possibly as part of unit V by Nathan (1969). Other enrichments on Little Iron Lake, mostly in unit T, were regarded by Nathan and subsequent students as having greater potential but were not investigated by Grout. Thus resources listed by Grout can only be suggestive of total resources of the area.

Grout (1949–1950) showed that an ilmenite fraction containing more than 40 percent TiO_2 could be made from some oxide-mineral enrichments of the area and that this fraction constituted as much as 12 percent of some rock bodies. I now regard only the 2.1 million metric tons of ilmenite listed by Grout as a demonstrated resource of separable ilmenite; the mineralogy and size of other enriched bodies included by others, and repeated by Force and Lynd (1984), are insufficiently known.

Pluma Hidalgo district, Mexico

The state of Oaxaca contains a large terrane of high-grade metamorphic rocks and structurally underlying anorthosites (Ortega-Gutierez, 1981) approximately 1,000 m.y. old. These rocks host rutile deposits near the town of Pluma Hidalgo, accessible only by a narrow road hung along a steep seaward-facing escarpment among rain forest and coffee plantations. Weathering is deep, except in road cuts and streambeds, so exposure of fresh rock is poor. The following description of the deposits is from my own observations, supplemented by Paulson (1964) and an unpublished (1957) description by T. P. Thayer. No adequate geologic maps of the district have been published.

The structural sequence is displayed in the face of the escarpment and shows gneissic country rocks toward the top and anorthosite at the bottom. The gneissic country rocks are mostly dark, finely banded quartz–antiperthite–two pyroxene–garnet–graphite–ilmenite gneiss, with numerous thin sills of quartz-feldspar rock. Banding is folded but remains subhorizontal over

large areas. A strong subhorizontal lineation trends northwest-southeast.

Passing structurally downsection, these gneisses are intruded by concordant and discordant bodies of impure anorthosite containing antiperthite and pyroxene megacrysts and quartz. These rocks host the Pluma Hidalgo deposits. Still farther downsection, toward the bed of the Toltepec River, anorthosite is massive but deformed and shows the same northwest-southeast lineation. It seems appropriate to call this body the Toltepec Anorthosite. Farther down toward the Pacific coast the anorthosite is highly altered with the addition of much silica and tourmaline.

In the immediate region of Pluma Hidalgo, small intrusions of the impure anorthosite are abundant and generally contain 1 to 2 percent coarse rutile. The rutile deposits of greatest economic interest, however, are impure anorthosites that contain from about 2 to more than 50 percent rutile (Fig. 32), probably averaging 20 percent by weight in the deposits known as Agua Titania or Las Minas de Tisur. These may be the highest-grade rutile deposits of any origin in the world. Rutile-bearing intervals average approximately 20 to 40 m in width over a strike length of at least 600 m, exposed in four perilous adits and in the steep bed of Agua Titania. Wall rocks of these zones are mostly gneiss but locally are impure anorthosite with lower rutile contents.

Rutile occurs as coarse single crystals; in high-grade samples, it has the appearance of a matrix between altered feldspar or pyroxene megacrysts. Ilmenite is also present in most specimens as coarse single crystals without intergrowths. In some high-grade material, low rutile content is compensated by high ilmenite content.

Some form of late- or postmagmatic mass transport of titanium must have occurred to form the high-grade rutile deposits of Pluma Hidalgo, as deposits of this type normally have lower rutile contents. The contact-metasomatic hypothesis fits the geologic relations but can be only one ingredient in the origin of these rutile deposits.

Kragerø district, Norway

Rutile was mined until 1927 from rutile-rich albitites near Kragerø on the southern coast of Norway. Probably several tens of thousands of tons of rutile were produced. The deposits are now nearly mined out, and today the region is a popular resort area that is part of a lovely coastal archipelago of glaciated islands (Fig. 33). Rock exposure is exceptionally good.

The deposits and their host rocks have been described by Brøgger (1934–1935) and Green (1956). Regional geology has been reviewed by Starmer (1985a) and mapped by Starmer (1985b). The immediate area of the deposits has been mapped by Green (1956).

Some aspects of regional geology in the district were described in a previous section of this chapter. Rutile albitites are present in two types of amphibolitic host rocks. Probably the more important type is foliated metagabbro containing scapolite and sphene. This type, though relatively homogeneous, locally

Figure 33. The Kragerø archipelago, photographed from the former rutile mine at Lindvikskollen, Norway, looking east to islands of amphibolite, quartzite, and gneiss. Langøy is in extreme upper left.

shows well-developed cumulate layering. Where gabbro bodies are little altered, these cumulates include magnetite bands rich in Ti and V. The other type of amphibolite is deformed and recrystallized pillow lava. Locally, as on western Langøy island, albitite preferentially replaces interpillow hyaloclastite (Fig. 34).

Rutile is most abundant in albitite where amphibolite country rocks are rich in sphene. Locally, the shape, size, and orientation of clots of sphene crystals in amphibolite are reflected by the distribution of clots of rutile, ilmenite, and sphene in crosscutting albitite bodies (Figs. 18, 35). Thus, metasomatism of country rock without TiO_2 transport is required in at least the marginal zones of albitite bodies. Green (1956) also reported this phenomenon.

The main rutile ore body at Kragerø was a zone about 2 m wide, parallel to the walls of a large concordant body of albitite containing 1 to 2 percent rutile. This zone averages 6 to 10 percent rutile but is quite inhomogeneous. The zone apparently represents a former train of xenoliths; the sizes and shapes of individual xenoliths are outlined by rutile-rich albitite (Fig. 36). The xenoliths differ from one another in texture and configuration. Some contain more than 25 percent rutile; others contain tourmaline or corundum and less rutile. The xenoliths include

many of equant angular shape, measuring up to 20 cm perpendicular to elongation of the zone. For this reason the term *xenolith* is more appropriate than *schlieren,* the term used by Brøgger (1934–1935).

The original lithology represented by these xenoliths is uncertain, as the train could not be traced into amphibolite country rock. Two possibilities are discrete pillows in amphibolite and xenolith trains in metagabbro. Since the metasomatic replacement of xenoliths by albitite is simplest if the precursor contained 10 to 30 percent TiO_2, I suggest that the precursor was a cumulate band of titaniferous magnetite in metagabbro, disrupted by the formation of albitite.

ECONOMIC PROGNOSIS

Magmatic ilmenite deposits in anorthosite-ferrodiorite massifs currently supply about 30 percent of the world's titanium minerals, and large high-grade resources remain. Objections to pollution from sulfate-process refining could make retention of this status difficult, but smelting and synthetic-rutile technologies, which can be used with these ores either singly or in tandem, greatly decrease pollution. Thus, magmatic ilmenite deposits

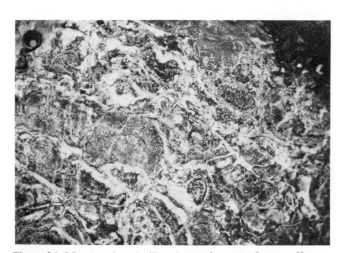

Figure 34. Metamorphosed pillow lavas of western Langøy, Kragerø district, with interpillow area replaced by albitite. Lens cap 50 mm across.

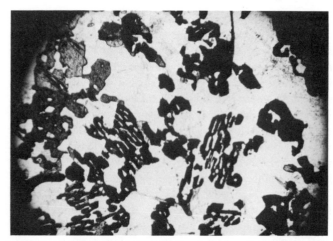

Figure 35. Photomicrograph of albitite from locality of Figure 18, Kragerø, showing skeletal ilmenite + quartz pseudomorphic after sphene (upper left). Transmitted plane light, 2-mm field.

Figure 36. Rutile ore exposed in pillar of Lindvikskollen mine, Kragerø. Host rock is light-colored albitite. Rutile-rich albitite is present as darker square-cornered ghosts of former xenoliths. The upper more shadowy xenoliths contain biotite, tourmaline, and locally corundum in addition to rutile.

probably will be able to compete with other deposit types for future world markets.

In the United States, these advanced technologies have not as yet been used to refine ilmenite from igneous deposits, whereas smelting is used in Canada and Norway. This situation is hard to explain, as some major ores in the United States are amenable to smelting (Elger and others, 1986; Nafziger and Elger, 1987). The economic future of magmatic ilmenite deposits in the United States depends less on its geology than on its technology.

Contact-metasomatic rutile deposits are normally low in grade; portions sufficiently high in grade to be economic are typically small. Therefore, these deposits will probably never

supply a significant share of world titanium markets; however, small-scale mining of some high-grade deposits may be profitable.

METHODS OF EXPLORATION

The most valuable magmatic ilmenite deposits occur in andesine anorthosite massifs associated with ferrodiorites; contact-metasomatic rutile deposits occur on the margins of alkalic andesine anorthosite massifs. Thus, exploration should focus on andesine anorthosite-ferrodiorite massifs. Labradorite anorthosite massifs and stratiform mafic complexes are not promising as sources of economic titanium minerals.

Magmatic ilmenite deposits tend to occur toward the bases of ferrodiorite sheets and/or in structurally underlying country rock within about 100 to 200 m of the reconstructed base of the overlying ferrodiorite sheet. Such deposits seem to be preferentially located in or under synclines in the ferrodiorite. Contact-metasomatic rutile deposits occur along contacts of alkalic andesine anorthosites with older country rock, especially along contacts where numerous dikes and sills of the anorthosite are intruded. Enrichments in TiO_2 to make such deposits economic may be present only at structurally and/or lithologically favorable sites.

Most of the deposits described here were discovered by physical exploration, but hidden extensions of magmatic ilmenite deposits have commonly been found by using aeromagnetic surveys. It would be far more efficient to start with the aeromagnetic surveys. In doing so, however, it must be borne in mind that the most valuable deposits contain ilmenite that is of a stoichiometric composition or is intergrown with hematite. Magnetite may be present only in the less attractive bodies, as in the ferrodiorite deposits of the Allard Lake district. Therefore, once within an area of magmatic ilmenite occurrence, all anomalies should be checked regardless of amplitude. Contact-metasomatic rutile deposits, of course, will show no magnetic anomalies.

Induced polarization methods were found by Elliot and Guilbert (1975) to respond to hemoilmenite ± magnetite concentrations. Magmatic ilmenite deposits consist in large part of electrically conductive grains in physical contact at least through the intercumulus spaces, so some conductivity method should be useful in their exploration.

Since mineralogy in both types of anorthosite-related titanium-mineral deposit determines the economic significance of a deposit, fluvial prospecting for ilmenite and rutile may be efficient in nonglaciated areas. For example, detrital ilmenite too coarse to be derived from gneisses could be sought and examined for favorable intergrowth type.

Chapter 4.

Titanium oxide minerals in alkalic igneous rocks

A second suite of igneous rocks that contains valuable titanium-mineral deposits is alkalic ring complexes of "miaskitic" type. The term *miaskitic,* introduced by Ussing (1911), refers to nepheline syenites having atomic alkali/alumina ratios of less than one; those having higher ratios are called *agpaitic.* Ussing pointed out that miaskitic nepheline syenites tend to occur in alkalic complexes in which high titanium and iron contents are present as oxide minerals. Thus the term *miaskitic* is applied to entire complexes containing miaskitic nepheline syenites. As we shall see, the two meanings of the term are not everywhere consistent. I will use the term rather loosely in the latter sense. A major class of miaskitic alkalic rocks occurs as ring complexes, with carbonatites typically forming the core unit and with accessory carbonate disseminated through the complex.

MINERALOGY

The important titanium-bearing oxide-mineral phases in miaskitic alkalic complexes, in order of probable abundance, are perovskite, magnetite, and the TiO_2 polymorphs rutile and brookite. These minerals, except for magnetite, commonly contain large amounts of niobium in alkalic rocks (Fleischer and others, 1952). Ilmenite is normally minor but may be abundant; indeed, the type locality of ilmenite is in a miaskitic intrusive of the Ilmen Range of the Ural Mountains, near Miask, U.S.S.R. (Sørensen, 1974).

In contrast, the agpaitic alkalic complexes typically contain their titanium mostly in silicates such as sphene, titanaugite, melanitic (or schorlomitic) garnet, kaersutitic amphibole, and several unusual Ti-Zr silicate minerals. Factors that suppress the formation of these silicates are responsible for the economic titanium deposits in miaskitic alkalic complexes. Possibly sphene is suppressed by some combination of the following bulk chemical variables in alkaline rocks: (1) low silica activity, which favors perovskite relative to sphene (Carmichael and others, 1970); (2) high contents of niobium, thorium, and rare earths, which can readily be accommodated in the lattices of perovskite, brookite, and rutile but are not reported in comparable concentrations in sphene; (3) high CO_2 content, which forms rutile plus calcite at the expense of sphene (Hunt and Kerrick, 1977); and (4) the low oxygen fugacity prevalent in alkalic rocks, which favors perovskite over sphene (Haggerty, 1976b).

TITANIUM-OXIDE MINERAL ENRICHMENTS

The most valuable enrichments of titanium oxide minerals occur in miaskitic complexes that contain pyroxenites and alkali pyroxenites (jacupirangites, etc.). Most such deposits are actually in the pyroxenites. In the Kola Peninsula of the U.S.S.R., enrichments of titanium oxide minerals occur in pyroxenites of both miaskitic (Africanda) and agpaitic (Khibiny, Lovozero) affinity based on nepheline syenite composition (Yudin and Zak, 1971). The agpaitic pyroxenites contain sphene in addition to magnetite and perovskite.

In many of the important deposits, the titanium-mineral enrichments occur as magnetite-perovskite rocks. These rocks are not well described in any deposit, but in my experience occur as small dike-like bodies permeating pyroxenite. Their origin has been attributed by Herz (1976) to an oxide liquid immiscible in alkalic magmas of intermediate composition. Relations described herein for the Kodal deposit (Norway) may represent an intermediate stage in the formation of such liquids.

An important titanium resource at Magnet Cove, Arkansas, is a contact-metamorphic deposit, formed where alkalic magma came in contact with silicic sedimentary country rock. Several deposits in alkalic rocks have a weathering overprint, and in Brazilian deposits the weathered overburden is the resource. Variations among deposit types are summarized in Table 10.

**TABLE 10. RELATION OF THE DESCRIBED TITANIUM-MINERAL DEPOSITS
IN ALKALIC IGNEOUS ROCKS**

Deposit	Deposits in igneous rocks			Contact-metamorphic deposits	Weathered deposits
	Pyroxenitic host	Syenite host	Other host		
Powderhorn district	X				
Brazilian deposits	X				X
Magnet Cove district Magnet Cove Rutile deposit			X		
Mo-Ti prospect	X				
Christy deposit				X	X
Hardy-Walsh deposit				X	X
Kodal deposit		X			

MAJOR DEPOSITS

Powderhorn (Iron Hill, Cebolla Creek) district, Colorado

Claims by the Buttes Oil and Gas Company (Wall Street Journal, 1976; Thompson, 1987) that the Powderhorn district contains about 500 million (short?) tons of perovskite ore imply that the district is the largest single titanium-mineral resource in the United States. However, beneficiation of perovskite to extract TiO_2 has not yet been demonstrated to be commercially feasible.

The district has been described by Larsen (1942), Heinrich (1966), and Temple and Grogan (1965) and mapped by Hedlund and Olson (1975). The titanium-mineral resources are in an alkalic ring complex with a carbonatite core in Precambrian crystalline rocks. It is part of a Cambrian swarm of alkalic intrusives in southern Colorado. The oxide-rich nature of the complex is not reflected in nepheline syenite composition (analysis in Larsen, 1942).

The Powderhorn area, at more than 8,000 ft (2,440 m) elevation, is on the northern flank of the San Juan Mountains and slopes toward the Gunnison River. The area is sparsely inhabited and forested. The pyroxenite unit, which composes about 70 percent of the 30-km^2 area of the complex, occupies a long intermontane depression (Fig. 37). In three dimensions the pyroxenite is cone-shaped (Temple and Grogan, 1965).

Pyroxenite consists of diopsidic augite, locally titaniferous, with lesser and varying amounts of magnetite, perovskite, biotite, and phlogopite. Minor calcite and apatite are ubiquitous. Important varieties of pyroxenite are defined by local abundance of apatite, olivine, nepheline, feldspar, and melanite garnet containing 5 percent TiO_2. Garnet is, like sphene, a late mineral and formed apparently at the expense of perovskite. Variations in both composition and texture make the pyroxenite an exceedingly inhomogeneous unit. Larsen (1942) considered the varieties mutually intrusive. The range in TiO_2 contents of pyroxenite is correspondingly great, 0.3 to 11.9 percent, averaging 6.5 percent (Rose and Shannon, 1960).

Magnetite-perovskite rock forms discontinuous dikes and lenses ranging in thickness from less than 0.5 m to almost 50 m (Hedlund and Olson, 1975). Abundant apatite defines one variety of this rock but normally is a minor constituent. Minor biotite is also present. Larsen (1942) reported perovskite contents as high as 50 percent and TiO_2 contents as high as 40 percent in magnetite-perovskite rock.

The petrology of this rock has not been adequately described; the following observations are my own. Perovskite forms crystals commonly 1 to 4 mm in diameter, some euhedral (Fig. 38). Magnetite contains coarse ilmenite lamellae. Ilmenite is also present as separate grains that grade in morphology into the lamellae in magnetite (Fig. 38B). Ilmenite locally rims perovskite.

Late microcrystalline sphene-anatase intergrowths (apparently the leucoxene reported in the literature) both rim perovskite and fill cracks in it. Some lithologies also contain apatite and clinopyroxene.

The resource figures by Rose and Shannon (1960) and by the Buttes Oil and Gas Company are based on estimates of the volume of pyroxenite that is unusually permeated by magnetite-perovskite veinlets; perovskite is thus treated as if it were disseminated. About 500 million or more tons of such pyroxenite average about 12 percent TiO_2 but contain only about 8 percent separable perovskite with 45 to 50 percent TiO_2, 1.5 percent rare earths, 0.5 percent Nb (Thompson, 1987, and oral communication, 1981), and appreciable thorium (Hedlund and Olson, 1975).

Tapira and Salitre (Minas Gerais State) and Catalão I (Goias State) deposits, Brazil

The anatase deposits in weathered overburden of alkalic bodies in Brazil are among the world's largest titanium-mineral resources; there is some industry speculation that these deposits could replace more traditional sources of titanium minerals (Industrial Minerals, 1978). Estimates of total resources are 300 million metric tons or more of anatase ore containing more than 20 percent TiO_2 (Mineração Metalurgia, 1977; Beurlen and Cassedanne, 1981; Turner, 1986) for the three bodies known as Tapira, Salitre, and Catalão I. Trade-journal articles furnish most of the available non-Portuguese information on these deposits (e.g., Harben, 1984; Turner, 1986), although some information on their parent alkalic bodies is available in older technical literature (Troger, 1928, 1935) and in more recent brief descriptions (Heinrich, 1966; Herz, 1976).

Ulbrich and Gomes (1981) list 44 alkalic stocks dating from 40 to 90 Ma, disposed around the Parana basin in Brazil. In addition to the three bodies containing titanium-mineral resources, other bodies such as Araxa contain important resources of rare earths, phosphate, baddeleyite, supergene manganese oxides, and garnierite. Some of the bodies are agpaitic, and numerous others are poorly known.

The three stocks of interest for titanium have been most thoroughly described by Alves (1960), Carvalho (1974), Filho (1974), and Geisel-Sobrinho (1974). All have been mapped by Barbosa and others (1970), who show them as circular stocks intruded into Precambrian rocks. Their surface areas range from about 25 to 40 km². Annotated air photos of each body, shown by Barbosa and others, indicate that the stocks are physiographically expressed as subtle domes largely mantled by residual debris. Country rock crystalline terranes show moderate relief and largely grassland vegetation. Annual rainfall averages 1.7 m (Harben, 1984), permitting deep weathering.

The alkalic stocks are not well known, in part because of their mantles of residuum. They seem to be sufficiently similar that they can be described in aggregate, based on references listed above. They all show marginal emplacement breccias and have

Figure 37. View of the Powderhorn district, Colorado, looking WSW. Pyroxenite occupies the grassy slopes beyond the wooded foreground of older Powderhorn Granite. Carbonatite occupies wooded Iron Hill in center. Photo courtesy of Theodore Armbrustmacher.

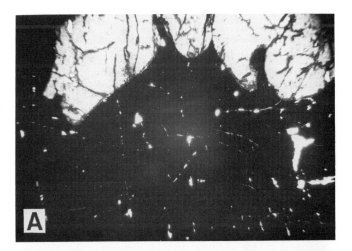

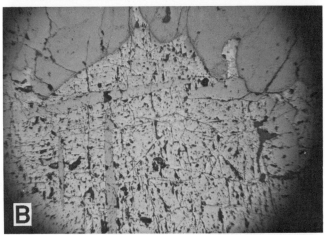

Figure 38. Photomicrographs of magnetite-perovskite rock, Powderhorn district. A. Transmitted plane light, 2-mm field. Perovskite, along the top, transmits light. B. Same field in reflected light, showing ilmenite in magnetite (pitted), both as separate grains and intergrowths.

Figure 39. Anatase pseudomorphic after perovskite and/or magnetite, Tapira, Brazil. Scale in millimeters.

small carbonatite cores. Bostonite (aplitic syenite) is present at Tapira. Jacupirangite, bebedourite, and other alkalic pyroxenites rich in oxide minerals are common to predominant lithologies in all three stocks and contain up to 14 percent perovskite and/or 30 percent sphene (Troger, 1935). Some less common rocks are further enriched in magnetite, perovskite, sphene, and apatite; in some, apatite is the most abundant mineral.

The petrography of magnetite-perovskite-sphene rocks at Tapira is illustrated by Alves (1960). All three minerals form coarse independent grains, commonly with apatite. Magnetite contains ilmenite lamellae and is partially replaced by hematite.

The potential ores of these deposits, however, are not the titaniferous alkalic rocks, but parts of the weathered residuum over them, which can be as thick as 200 m. Anatase (octahedrite) in this residuum is microcrystalline and replaces perovskite, magnetite, and sphene. Anatase may be pseudomorphic after octahedral perovskite and/or magnetite (Fig. 39); that is, the octahedral form of anatase masses is from other octahedral minerals. Anatase masses vary in size from more than 1 cm to grains less than 0.1 mm in diameter. Some masses in my specimens have cores of primary minerals such as perovskite. The microcrystalline anatase is porous and cemented by iron hydroxides (Turner, 1986).

Weathered residuum of the three stocks contains phosphate resources in addition to titanium-mineral resources. The weathering profile shows a basal zone rich in apatite, an intermediate anatase- and magnetite-rich zone, and more barren surficial material (Cruz and others, 1977; Harben, 1984). At Tapira, the anatase ore of the intermediate zone is 40 to 55 m thick and contains more than 20 percent TiO_2. Titanium is present in magnetite, perovskite, and Ti-rich schorlomitic garnet as well as anatase (Harben, 1984). Herz (1976) believes that more than 60 percent of the TiO_2 is present as anatase, but no substantiation for this figure appears in the literature. Thus there is currently no good

basis for calculating reserve figures for these bodies from published data on tonnage and TiO_2 content.

Magnet Cove district, Arkansas

A small alkalic ring complex of Mesozoic age, intruded into folded Paleozoic sedimentary rocks, forms the Magnet Cove district near Hot Springs, Arkansas (Fig. 40). The alkalic igneous rocks form a basin, or cove, which interrupts the Zigzag Mountains of sedimentary rocks. Titanium-mineral production from the Magnet Cove area has been minor.

The geology and geochemistry of the district have been described by Erickson and Blade (1963), and the district has been mapped by Erickson and Blade and by Danilchik and Haley (1964). Compositions of nepheline syenite (Tables 3 and 7 of Erickson and Blade) show both miaskitic and agpaitic values. Erickson and Blade relate their 28 alkalic lithologies to three rings; the innermost contains carbonatite. The igneous hosts for titanium-mineral deposits are jacupirangite of the outer ring and altered phonolite of the intermediate ring. In addition, garnet nepheline syenite and jacupirangite of the outer ring may be related to a contact-metamorphic deposit of brookite in altered sedimentary country rocks.

Jacupirangite, or magnetite-perovskite pyroxenite, contains 4.0 to 4.3 percent TiO_2, present in the following titaniferous phases: titaniferous clinopyroxene (78 percent average mode) containing 3.0 percent TiO_2, perovskite (4 modal percent), minor titaniferous melanite garnet, and an undetermined amount of magnetite intergrown with ilmenite (Erickson and Blade, 1963). Apatite and calcite are also present. Near contacts with siliceous Arkansas novaculite wallrock, late sphene takes the place of perovskite and magnetite.

Altered phonolite, locally brecciated, averages 2.5 percent TiO_2 and contains the Ti-bearing phases magnetite-ilmenite and sphene. Abundant late calcite and minor apatite are present. Garnet-nepheline syenite contains 0.8 to 1.1 percent TiO_2. Titaniferous phases are zoned melanitic garnet (as much as 6 modal percent), which in this district contains up to 15 percent TiO_2, titaniferous clinopyroxene, and minor magnetite and sphene.

The Magnet Cove alkalic intrusives are unusual in showing little fenitization of country rocks (Heinrich, 1966). Thus, alkalic igneous rocks impinge directly on sedimentary rocks. Flohr and Ross (1989) note fenitized xenoliths at one locality.

Titanium-mineral deposits of the district have been exhaustively described. In all, their treatment by Kinney (1949), Reed (1949a and 1949b), Fryklund and Holbrook (1950), Toewe and others (1971), and many others comes to more than 450 pages, plus maps. The identified resources, however, appear to be of modest economic importance, as tonnages are small and mineral compositions unfavorable. The following descriptions are from those sources.

The deposits occur in three main settings: rutile in feldspar-carbonate veins in altered phonolite, brookite in feldspar-pyrite

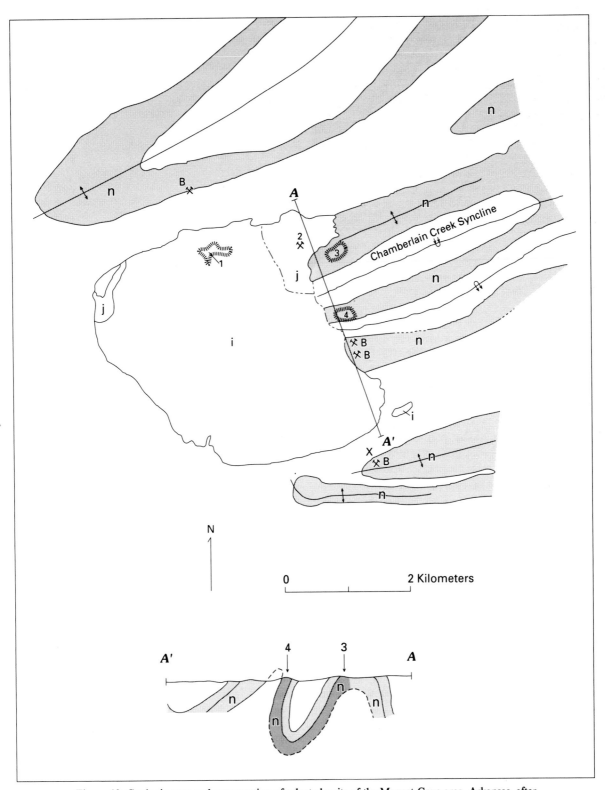

Figure 40. Geologic map and cross section of selected units of the Magnet Cove area, Arkansas, after Erickson and Blade (1963) and Danilchik and Haley (1964). Deposits are 1, Magnet Cove Rutile; 2, Mo-Ti; 3, Hardy-Walsh; and 4, Christy. Other brookite prospects labeled B. See text for locality X. Unit n, novaculite; i, alkalic igneous rock, including j, jacupirangite. See text for locality X and resources of dark-shaded unit in cross section.

Figure 41. Network of quartz-brookite-limonite veinlets (dark) in quartzite, Christy deposit, Magnet Cove. All units weathered; exposure about 4 m high.

veins in jacupirangite, and brookite in skarn-like (or fenite-like?) contact metamorphic deposits.

Deposits in igneous rocks. The Magnet Cove Rutile deposit (Fig. 40) produced minor amounts of rutile from 1932 to 1944. Rutile occurs in swarms of feldspar-carbonate veins that cut porphyritic aegirine phonolite and a variety of alkalic dike rocks, all hydrothermally altered in places. The rutile-bearing veins contain albite or microcline; dolomite, calcite, or ankerite; and pyrite. Rutile occurs as prisms averaging 0.1 mm in length, locally aggregated into irregular masses. The rutile content of intervals as long as 0.6 m is up to 5 percent and averages 2.7 percent. The rutile contains 1 to 2 percent niobium and 0 to 1 percent V_2O_5, and these impurities make it difficult to use. Resources are not thought to be large anyway.

The Mo-Ti prospect (Fig. 40) contains brookite, and pyrite coated by molybdenite, in microcline veins through a large jacupirangite mass. The total mineralized area measures at most 10 m x 130 m (Holbrook, 1948), so resources are thought to be minor. Perovskite containing 0.6 percent niobium is reported from the host rock.

Contact-metamorphic deposits. The Christy and Hardy-Walsh (Kilpatrick) brookite deposits formed at the eastern margin of the igneous complex in the lower member of siliceous Arkansas novaculite, where it is in contact with alkalic igneous rocks (Fig. 40). The lower member is about 100 m thick, and the deposits are within about 300 m of intrusive rocks. Igneous rock nearest the Christy deposit is garnet nepheline phonolite, whereas that nearest the Hardy-Walsh deposit is jacupirangite. Both deposits are weathered. The average abundance of recoverable brookite is about 5 percent. The brookite contains 2 percent Nb and 0.5 percent V_2O_5. The Christy deposit is presently being mined for vanadium.

Both deposits consist of a network (Fig. 41) of quartz-

brookite-limonite veinlets and disseminations in quartzite recrystallized from novaculite. Open spaces have permitted the formation of euhedral crystals of several minerals, including brookite. Taeniolite ($KLiMg_2Si_4O_{10}F_2$) mica occurs as disseminations and irregular masses, and increases westward toward the intrusives. Fine rutile is a minor constituent. Brookite averages 0.5 to 1 mm in diameter and reaches 6 to 7 mm. It is slightly enriched in clayey weathered zones about 5 m thick. Where truly unoxidized rock was encountered by Fryklund and Holbrook (1950), pyrite was present in place of limonite.

Apparently, titanium in these deposits has been introduced from TiO_2-rich intrusive rocks, such as jacupirangite, into novaculite that contains almost no TiO_2 (Cressman, 1962, Table 6). Other elements apparently introduced from the intrusive were Fe, V, Li, F, Nb, and perhaps S. Elements moving into the intrusive included Si (to form jacupirangite with secondary sphene). Thus these contact-metamorphic deposits suggest some unusual elemental fluxes. In normal skarn deposits, titanium is not mobile; indeed, titanium distribution is sometimes used to delimit the original extent of igneous rock (cf., Large, 1972).

Six brookite prospects have the same geologic settings as the Christy and Hardy-Walsh deposits (Fig. 40). Their distribution suggests to me that the total resources of this deposit type at Magnet Cove should be calculated as all coarse brookite present in the lower member of novaculite (100 m thick) between 100 and 300 m from the alkalic complex, through the Chamberlain Creek syncline and adjacent structures. The cross section (Fig. 40) presents this view of resource potential, using a depth to novaculite at the syncline axis (890 m) calculated from accurate data on the plunge in the adjacent barite mine (Scull, 1958). The implied brookite resource in the Chamberlain Creek syncline alone is on the order of 5×10^6 metric tons. This is a hypothetical resource of respectable magnitude, worthy of further investigation from the points of view of industrial utility of this brookite, corollary vanadium resources, and further exploration. For example, drilling to the lower member of novaculite at location X in Figure 40 could be warranted.

Kodal deposit, Norway

The Oslo alkalic igneous province contains syenitic composite ring intrusions of miaskitic affinity (from analyses of nepheline syenite in Oftedahl, 1960). These syenites are called larvikite (augite syenite) and lardalite (nepheline syenite). The Kodal deposit is associated with a nepheline larvikite ring showing rhythmic igneous banding (Petersen, 1978; Lindberg, 1986). It can be described either as a jacupirangite or as an impure nelsonite (Bergstøl, 1972); I refer to it here as jacupirangite.

The main jacupirangite body dips steeply south, is 1,900 m long, and varies in thickness from 2 to 32 m (Nielsen, 1972). In addition to pyroxene, it contains 17 percent apatite, 40 percent magnetite, and from 5 to 15 percent ilmenite (averaging 8 to 9 percent). Ilmenite and magnetite form separate grains 0.1 to

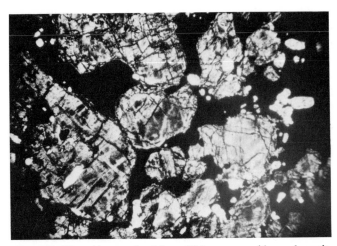

Figure 42. Photomicrograph of interstitial opaque oxides and apatite among cumulate augite crystals, ore band at Kodal, Norway. Transmitted plane light, 6-mm field.

1.0 mm in diameter. Ilmenite grains contain rutile and magnetite intergrowths, and magnetite grains contain ingrown ilmenite. Apatite is finer grained. Lindberg (1986) describes sharp footwall contacts below which is a 100-m zone of larvikite impregnated by oxide minerals and apatite. The hanging wall is gradational, and the overlying larvikite contains numerous stratiform lenses enriched in oxide minerals and apatite.

As noted by Petersen (1978) and Lindberg (1986), the jacupirangite body itself shows cumulate bands outlined by modal and grain-size variations. Oxide minerals and apatite are interstitial to pyroxene in jacupirangite (Fig. 42; cf., Bergstøl, 1972) and in the underlying impregnated zone (my observation). Locally, pyroxene cumulates contain macroscopic blebs of oxide minerals and apatite in a range of sizes. This evidence suggests immiscibility of a nelsonitic liquid in larvikitic magma, by the same reasoning developed in Chapter 3. Similar conclusions were reached by Bergstøl (1972) and Lindberg (oral communication, 1987).

The Kodal deposit is less alkalic than others described in this chapter. Its oxide mineralogy (magnetite, ilmenite with intergrown rutile) and mode of formation (cumulates of dense immiscible liquids) suggest a transition to ilmenite deposits of anorthosite-ferrodiorite massifs.

ECONOMIC PROGNOSIS

Titanium-oxide mineral deposits in alkalic rocks suffer from a great economic disadvantage compared with deposits of other types: their oxide minerals contain niobium and, commonly, rare earth elements and thorium. This makes refinement difficult. Some deposits contain perovskite, for which TiO_2-recovery technology is still uncertain. Deposits like Kodal are dominated by magnetite, which has no value as a titanium mineral.

Some deposits of titanium oxide minerals in alkalic rocks also are too small, for example, the Magnet Cove Rutile deposit. One type of deposit that suffers from an unfavorable mineral composition but not from small size is the contact-metamorphic type, of which the Christy deposit is an example.

Titanium-oxide mineral deposits in weathered overburden over titanium-rich alkalic rocks have neither disadvantage, and these deposits are thus placed most favorably for commercial competition. The titanium-oxide mineral assemblage of these deposits is reconstituted in the weathering zone and thereby is mostly purged of detrimental trace elements and calcium, enabling treatment by the chloride process. The volume of some weathered overburdens of this type, such as those in Brazil, rivals those of large magmatic ilmenite deposits. Thus, these deposits are in a good position to capture a significant share of world titanium markets.

METHODS OF EXPLORATION

The formation of valuable anatase deposits in weathered alkalic rocks is clearly a two-stage process. The second stage, weathering, is not discussed in this chapter but in chapter 6. Here we need to discuss exploration for alkalic stocks rich in titanium oxide minerals. Location of promising stocks may be aided by their distinctive shapes in air photos such as those shown by Barbosa and others (1970), by soils maps if available, and by aeromagnetic surveying, as magnetite invariably seems to accompany perovskite or ilmenite in large Ti-rich bodies. If fresh rock is exposed, selection of the more favorable stocks can be on the basis of miaskitic composition of nepheline syenites; that is, the ratio Na_2O+K_2O/Al_2O_3 should be less than one (on an atomic basis); however, this has proved an unreliable indication in several districts. Heavy mineral concentrates could delineate subareas containing a titanium mineralogy suitable for upgrading by weathering in the regolith.

Chapter 5.

Titanium oxide minerals in other igneous suites

Titanium oxide minerals show interesting distribution patterns, but form no economic deposits, in igneous suites other than anorthosite-ferrodiorite massifs and alkalic complexes. The hydrothermal alteration of hypabyssal granitoid systems forms fine rutile that is of some economic interest.

PRIMARY TITANIUM OXIDE MINERALS OF SOME IGNEOUS SUITES

Titanium-mineral assemblages of primary and high-temperature oxidation origin are summarized here for four igneous suites: two igneous suites, the granitic and basaltic rocks, are selected for discussion because of their great volume and to correct misleading statements about their rutile contents (reviewed by Force, 1980b). Two less voluminous suites, layered mafic intrusives and kimberlites, are included because their oxide minerals are the subject of a voluminous literature. These discussions rely heavily on two review monographs by Haggerty (1976a, b), remarkable for their 107 pages of tables and 185 photomicrographs. An additional review is by Elsdon (1975). Partitioning analyses between oxides and silicates are mostly from Force (1976a).

Granitoid rocks

The TiO_2 contents of granitoid rocks range from about 0.3 to 0.7 percent; the high values are in lithologies of more intermediate composition. Silicate minerals typically contain 60 to 95 percent of this TiO_2. However, variations in partitioning of TiO_2 among oxide and silicate minerals are among the criteria used to classify granites. The I-type of Chappell and White (1974) or magnetite series of Ishihara (1977) contains 0.1 to 2 modal percent magnetite that contains ilmenite as lamellae, accompanied by up to 1 percent sphene, and Ti-poor hornblende. The S-type or ilmenite series contains only 0 to 0.2 percent ilmenite, accom-

panied by biotite. Haggerty (1976a) points out that the high oxygen fugacity represented by the I-type or magnetite granites favors oxides over silicate minerals, but that the oxide minerals represent the Fe-rich rather than the Ti-rich members of their respective solid-solution series.

Haggerty (1976a, Table Hgl2) shows that typical primary oxide-mineral assemblages in granitoid rocks are coexisting ulvospinel-magnetite and ilmenite-hematite, each showing exsolution textures. Exsolution oxidation of ulvospinel to ilmenite is common, and later pseudomorphic oxidation to intergrown hematite and rutile is known. The MnO contents of ilmenite in granitoid rocks are high, and especially in peralkaline granites this manganoan ilmenite can properly be called pyrophanite.

Reports of titanium-mineral concentrations in granitoid rocks are few. Nelsonite-like concentrations of titaniferous magnetite and apatite in breccia matrix and dikelike bodies occur in syenitic hypabyssal and other granitoid stocks of the Canadian Cordillera (Badham and Morton, 1976; W. J. McMillan, written communication, 1981). Some of these have been explained as immiscible liquids.

The occurrence of rutile as discrete grains in granitoid rocks and related pegmatites has been reviewed by Force (1980b). Granular accessory rutile is reported from very few granitoid rocks (Lee and Dodge, 1964), although secondary rutile included in other oxide-mineral grains is fairly common. In some granites from which rutile has been reported, such as those of Dartmoor and Cornwall, Great Britain (Bramall, 1928; Groves, 1931), rutile is present in late veins that accompany hydrothermal alteration, commonly as pseudomorphs of titaniferous biotite (D. R. Wones, oral communication, 1980).

Niobian rutile (ilmenorutile) has been reported from hypabyssal leucocratic stocks bearing molybdenum and tin mineralization (Sainsbury, 1968; Desborough and Sharp, 1978; Desborough and Mihalik, 1980). This rutile is apparently of primary origin, as it forms discrete grains in unaltered rock.

TABLE 11. TITANIUM OXIDE MINERALS OF SOME IGNEOUS SUITES OTHER THAN THOSE DISCUSSED IN TEXT*

Rock Suite	TiO₂ Range (%)	Primary Mineralogy[†] Mag-Usp	Primary Mineralogy[†] Hem-Ilm	Exsolution Oxidation[†]	Pseudomorphic Oxidation[†]
Rhyolitic rocks	0.2–0.6	P	P	ilm (R)	hem (R)
Andesites	0.8–2.6	P	P	ilm (C)	hem, mgh (C)
Trachybasalts	2.6–2.8	P	P	ilm (C)	R
Syenites	0.7–1.1	P	P	R	R
Alpine mafic	0.02–1.0	R	R		

*From Haggerty, 1976a, and Force, 1976a.
[†]Abbreviations: Mag = magnetite; Usp = ulvospinel; Hem = hematite; Ilm = ilmenite; Mgh = maghemite; P = predominant; C = common; R = rare.

Basaltic rocks

The TiO₂ contents of typical basaltic rocks range from 1.5 to 2.7 percent. Partitioning is poorly documented, but calculations from partial analyses show that oxide minerals commonly contain more than 50 percent of this TiO₂. For example, figures in Gottfried and others (1968) show that oxide minerals contain 65 percent of the TiO₂ in early diabases, 86 percent in Fe-Ti–rich pegmatitic facies, and 47 percent in granophyres of the Dillsburg diabase sill of Pennsylvania. The variation is due in part to crystallization of ilmenite, which typically begins midway through the crystallization history of basaltic rocks (Wright and Peck, 1978).

Primary oxide minerals normally include members of both the ulvospinel-magnetite and ilmenite-hematite series, both in tholeiitic and alkaline olivine basalt types (Haggerty, 1976b). Exsolution oxidation to ilmenite is common, though possibly less common in alkaline olivine basalt. Further pseudomorphic oxidation to rutile, hematite, and/or pseudobrookite is widespread except in deep-sea basalts and is more advanced in flow centers than on flow margins. Granular accessory rutile in basalt is unknown.

Layered mafic intrusions

Stratiform, or layered, mafic intrusion complexes may contain layers enriched in iron-titanium oxide minerals, typically of cumulate origin. These layers contain cumulate plagioclase, clinopyroxene, and olivine and occur in the upper parts of complexes with ultramafic basal lithologies. Magnetite commonly dominates over ilmenite and contains exsolved ulvospinel. Locally, oxidation of ulvospinel forms ilmenite. Primary ilmenite contains exsolved hematite. Stratigraphically lower layers typically contain oxide minerals that are richer in iron in each solid-solution series. These lower layers may also have higher ratios of the magnetite series to the ilmenite series (reviewed by Haggerty, 1976a, from Skaergaard, Kiglapait, Muskox, Kap Edward Holm, Somerset Dam, and Sudbury). Chromite-bearing layers locally contain rutile of probable cumulate origin (Cameron, 1979).

In general, titanium-oxide mineral accumulation in layered mafic intrusions seems to begin at a later stage than in ferrodiorites of anorthosite massifs, probably because of liquid immiscibility in the latter. Locally, however, massive nelsonitic rocks consisting of magnetite (with intergrown ilmenite) and apatite may form discordant to concordant bodies attributed to immiscible liquids in layered complexes (Grobler and Whitfield, 1970).

Kimberlites

The titanium oxide minerals of kimberlites have been intensively investigated as clues to the composition of the lower crust and mantle and as prospecting tools in diamond exploration. The oxide mineralogy, which is quite complex because of lack of equilibrium, has been reviewed by Haggerty (1976a), Pasteris (1980), and Mitchell (1986). Ilmenite (as much as 10 percent), rutile, and perovskite are common in kimberlites, and armalcolite [(Fe,Mg)Ti₂O₅] has been reported. Both ilmenite and rutile occur as megacrysts, in the groundmass, and in intergrowths with other minerals. Ilmenite is characteristically rich in Mg and Cr and is called picro-ilmenite (ilmenite-geikielite_ss). Magnesium contents of ilmenite tend to be higher in the groundmass than in megacrysts.

Other suites

Haggerty (1976a) also gives information on titanium oxide minerals of other igneous suites, which is summarized in Table 11. Correlative partitioning analyses are not currently available in the literature; a rigorous study of titanium partitioning between oxide and silicate minerals is needed.

RUTILE IN HYDROTHERMAL ALTERATION ASSEMBLAGES

An extensive literature now exists on rutile associated with the hypabyssal granitoid rocks of porphyry alteration systems

TABLE 12. RUTILE-PRODUCING REACTIONS IN PORPHYRY SYSTEMS

Biotite + sulfur $-\,-\,-\,\blacktriangleright$ phlogopitic biotite + pyrite + rutile

$$K(Fe,Mg)_2(Fe,Al,Ti)Si_3AlO_{10}(OH,F)_2 + S_2 \,-\,-\,\blacktriangleright$$
$$K(Mg,Fe)_3 Si_3AlO_{10}(OH,F)_2 + FeS_2 + TiO_2$$

Hornblende + sulfur $-\,-\,\blacktriangleright$ actinolitic hornblende + pyrite + rutile + alumina

$$Ca_2(Fe,Mg,Ti,Al)_5Si_7AlO_{22}(OH,F)_2 + S_2 \,-\,-\,-\,\blacktriangleright$$
$$Ca_2(Mg,Fe)_5Si_8O_{22}(OH,F)_2 + FeS_2 + TiO_2 + Al_2O_3$$

Titaniferous magnetite + sulfur $-\,-\,-\,-\,\blacktriangleright$ lesser magnetite + pyrite + rutile

$$2(Fe,Ti)_3O_4 + S_2 \,-\,-\,-\,-\,\blacktriangleright Fe_3O_4 + FeS_2 + TiO_2$$

Ilmenite + sulfur $-\,-\,-\,-\,\blacktriangleright$ pyrite + rutile

$$FeTiO_3 + S_2 \,-\,-\,-\,\blacktriangleright FeS_2 + TiO_2$$

Sphene + carbon dioxide $-\,-\,-\,-\,\blacktriangleright$ rutile + calcite + quartz

$$CaTiSiO_5 + CO_2 \,-\,-\,-\,-\,\blacktriangleright TiO_2 + CaCO_3 + SiO_2$$

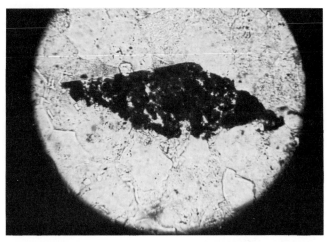

Figure 43. Photomicrograph of rutile (dark, high relief) + carbonate + quartz pseudomorphic after sphene, from propylitically altered rocks at Ajo, Arizona. Transmitted plane light, 6-mm field.

(Lawrence and Savage, 1975; Williams and Cesbron, 1977; Force, 1980a; Czamanske and others, 1981; Udubasa, 1982; Force and others, 1984). Rutile is a secondary mineral, coeval with hydrothermal alteration, and its formation is virtually inevitable if this alteration is sufficiently severe. The following discussion of rutile formation mostly follows Czamanske and others (1981).

Process of formation

Rutile in porphyry alteration systems records the amount and position of titanium present in fresh parent rocks, which are commonly calc-alkaline granitoids containing 0.3 to 1.0 percent TiO_2. Titaniferous minerals in these parent igneous rocks are magmatic biotite, magmatic hornblende, titaniferous magnetite, sphene, and ilmenite. Any one of these minerals may be the most important carrier of titanium in a given intrusion. None of these minerals is stable in the alteration environment, however; biotite, hornblende, and magnetite alter incongruently to new minerals containing less TiO_2, and ilmenite and sphene disappear. Table 12 shows some of the apparent mineral reactions, largely driven by the introduction of sulfur and CO_2. Figures 43 and 44 show some pseudomorphic relations that suggest these reactions. Table 13 contrasts the initial magmatic and subsequent post-alteration TiO_2 contents of biotite and amphibole.

Studies to date indicate that liberation of titanium is most complete, and resultant rutile of coarsest grain size, in the most altered zones of the alteration system—the potassic (biotite, K-feldspar) zone at San Manuel, Arizona, and Bingham, Utah, and the advanced argillic (andalusite) zone at Tangse, Sumatra (Force and others, 1984). Commonly these zones also have the richest or

"highest grade" sulfide mineralization. There is some evidence that phyllic (quartz-sericite) overprinting on higher grade zones does not destroy previously formed rutile. In propylitic (chlorite-epidote-carbonate) zones, rutile is very fine grained or absent.

The amount of rutile formed varies with the titanium content of the parent rocks, as titanium abundance is apparently conserved during alteration. Deposits developed on two igneous parents having differing TiO_2 contents and mineralogy have consequent differences in rutile content and morphology.

Bingham, Utah, example

Rutile has been recorded in ten porphyry deposits in the United States. Rutile occurrence at the Bingham Canyon deposit of Utah is described here as an example, because it is best known and possibly is the most valuable rutile resource. Bingham, until recently the largest U.S. producer of copper, is a copper porphyry with subordinate molybdenite. The deposit is in Tertiary intrusions of mesocratic equigranular quartz monzonite (MEQM) and later leucocratic porphyritic quartz monzonite (LPQM) in Paleozoic sedimentary rocks (Bray and Wilson, 1975, and maps therein). The ore is approximately coincident with potassic (biotite, K-feldspar) alteration. Propylitically altered and unaltered MEQM is present outside the potassic zone, and sericitic alteration is overprinted on potassic alteration in both intrusions. MEQM averages 0.9 percent TiO_2 and 15 to 25 percent biotite, whereas LPQM contains 0.6 percent TiO_2 and less than 10 percent biotite. The TiO_2-bearing minerals in unaltered MEQM include minor ilmenite and Ti-poor magnetite in addition to biotite (G. K. Czamanske, written communication, 1979). The difference in TiO_2 content between magmatic biotite and phlogopitic

E. R. Force

**TABLE 13. CONTRAST OF TiO$_2$ CONTENT OF
MAGMATIC AND HYDROTHERMAL BIOTITES AND
AMPHIBOLES IN PORPHYRY SYSTEMS**

	TiO$_2$ Content	
	Magmatic	Hydrothermal
Biotite		
Bingham, Utah*	4.925	1.65
Copper Canyon, Nevada[†]	4.7	1.5
Santa Rita, New Mexico[†]	4.1–4.7	1.5–3.1
Babine Lake, British Columbia[§]	4.3	2.8
Butte, Montana[†]	4.58	1.99
Amphibole		
Babine Lake, British Columbia[§]	2.0	0.4

*Moore and Czamanske, 1973.
[†]Czamanske and others, 1981.
[§]Carson and Jambor, 1974.

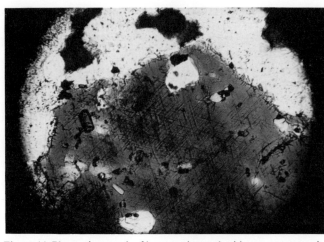

Figure 44. Photomicrograph of leucocratic porphyritic quartz monzonite from Bingham Canyon, Utah, showing rutile of two grain sizes in hydrothermal biotite after euhedral magmatic biotite—reticulate rutile needles and stubby prism (high relief, left center). Transmitted plane light, 2-mm field.

hydrothermal biotite (Table 13) averages about 3.3 percent TiO$_2$. Thus in the MEQM unit, 0.81 percent TiO$_2$ out of 0.9 percent TiO$_2$ in the rock is available to form rutile.

Czamanske and others (1981) found that rutile in MEQM and LPQM averages 0.34 and 0.24 weight percent, respectively, in the potassic alteration zone. The ubiquitous sagenitic rutile needles in altered magmatic biotite were not included. The average grain size of the coarser rutile was 60 x 30 μm in MEQM, 60 x 40 μm in LPQM. Trace-element impurities in this rutile total less than 1.5 percent (Czamanske and others, 1981).

Use of rutile in exploration

Lawrence and Savage (1975) and Williams and Cesbron (1977) suggested the use of rutile in exploration for porphyry deposits. Rutile is stable in the weathering environment (Chapter 6); of the entire hydrothermal mineral assemblage in a porphyry deposit, rutile is probably least susceptible to weathering. Force and others (1984) tested an exploration method based on rutile in soil samples. The presence of rutile in the regolith, and its grain size, were found to be potent indicators of alteration zones and sulfide distribution in unweathered underlying rock.

Economic prognosis

Rutile forms very large but very low-grade deposits in porphyry systems. It can be recovered only as a byproduct of mining for sulfides from these deposits. This rutile recovery has been evaluated most recently by Rampacek (1982) and Sillitoe (1983). The potential magnitude of rutile production is great.

Rutile is fine grained in porphyry deposits; less than half the rutile is as coarse as fine sand. Preliminary attempts using flotation and gravimetric methods to separate rutile from porphyry copper ores and tailings in the United States have been moderately successful (Llewellyn and Sullivan, 1980; Sullivan and Llewellyn, 1981; Davis and others, 1988). Such attempts have also been made with Chilean porphyry copper ores (Centro de Investigacion Minera y Metalurgia, 1986).

Chapter 6.

The weathering of titanium oxide minerals and the role of climate change

The weathering and erosion of titanium oxide minerals is the bridge between "primary" (igneous and metamorphic) and "secondary" (sedimentary) titanium-mineral deposits. This chapter treats the weathering not only of such crystalline rocks but also of sedimentary deposits and rocks. For this purpose, these sediments are treated solely as weathered rock; the following chapters discuss the sedimentary processes of titanium-mineral accumulation.

ECONOMIC IMPORTANCE

Weathering exerts a powerful influence on the relative compositions of titanium oxide–bearing rock, its weathered mantle, and its erosional debris. This influence is critical to the economic value of deposits, both in weathered primary rock and in sedimentary concentrations.

Titanium oxide minerals are mostly resistant to weathering, but they may be residually enriched in TiO_2. Weathering thus beneficiates titanium-mineral deposits in the following ways (Fig. 45): (1) Reduction in mass of the parent rock by leaching, at approximately constant volume. Normally this reduction is less than 50 percent, but greater reductions from calcareous and/or feldspathic rocks, locally to lesser volumes, are possible. Titanium oxide minerals are thus residually enriched. (2) The corollary destruction of such minerals as garnet, amphiboles, and pyroxenes, which have no economic value but are difficult to separate from titanium oxide minerals. (3) Disaggregation of the rock into monomineralic grains. This is economically important even in unconsolidated sediments, because detrital rock fragments may include both valuable and valueless minerals. (4) The chemical enrichment of TiO_2 in several titanium oxide minerals. This is particularly important for ilmenite, from which iron is leached in weathering. Other titanium oxide minerals may be leached of other ions; for example, perovskite may be leached of calcium to form microcrystalline anatase.

Weathering processes have upgraded a large number of titanium oxide mineral deposits and have made some of them economic. In a number of deposits discussed in previous chapters, the most valuable material is a saprolite developed on the primary deposit (Table 14). All detrital deposits can be regarded as upgraded by predepositional weathering. Superimposed in situ (post depositional) weathering of detrital deposits is also of great importance in some deposits.

THE BEHAVIOR OF TITANIUM IN THE WEATHERING ENVIRONMENT

Titanium as a chemical entity is residually enriched by weathering, as the more soluble elements are leached out of saprolites. Commonly this enrichment is less than 50 percent (Wells, 1960; Short, 1961; Dennen and Anderson, 1962; Loughman, 1969). However, enrichments in titanium by more than 100 percent have been reported even in noncalcareous rocks, where weathering is or was once intense (Goldich, 1938; Wahlstrom, 1948; Valeton, 1972; Patterson and others, 1986). In extreme weathering, TiO_2 may be mobilized within the rock to form concretions (Sherman, 1952).

Silicate minerals

Biotite, hornblende, titanaugite, sphene, and other titanium-bearing silicate minerals are among the less stable minerals in the weathering environment (Goldich, 1938; Dryden and Dryden, 1946). When they break down, their contained TiO_2 generally remains and commonly forms exceedingly fine-grained TiO_2 minerals such as anatase. Aggregates of these TiO_2 minerals may form pseudomorphs of the precursor silicates (Hartman, 1959; Valeton, 1972).

The rate of weathering varies with climatic conditions at the

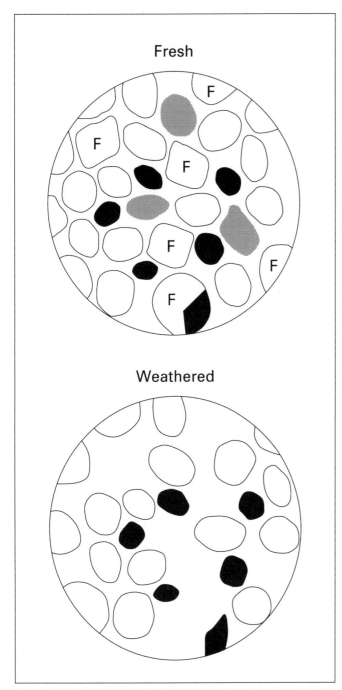

Fresh

Weathered

Figure 45. Diagram contrasting fresh and weathered rock (sand in this case) containing titanium oxide minerals. F, feldspar; shaded, pyroxenes and amphiboles; black, titanium oxide minerals; remainder quartz.

TABLE 14. TITANIUM OXIDE-MINERAL DEPOSITS KNOWN TO BE GREATLY BENEFICIATED BY WEATHERING

	Reference
Igneous and metamorphic deposits	
Roseland, Virginia	Herz and Force, 1987
Magnet Cove, Arkansas	Fryklund and Holbrook, 1950
Tapira, Salitre, and Catalão I, Brazil	Harben, 1984
Many rutile-bearing porphyry copper deposits	Czamanske and others, 1981
Postdepositional weathering of sedimentary placer deposits	
Trail Ridge, Florida	Pirkle and Yoho, 1970
Yoganup, Australia	Welch and others, 1975
Gbangbama, Sierra Leone	Raufuss, 1973
Lakehurst, New Jersey	Puffer and Cousminer, 1982
Predepositional weathering of detrital grains of placer deposits	
All placer deposits	Gillson, 1959

Residual enrichment of titanium oxide minerals

Rutile is among the most stable minerals in the weathering environment (Fish, 1962; Hubert, 1962; Overstreet and others, 1963; Force, 1976b). Microcrystalline rutile commonly forms therein from the weathering of ilmenite. Anatase, a polymorph of rutile, also forms in the weathering environment, mostly from silicate minerals in the parent rocks. The polymorph brookite is at least moderately stable, as it is found in the saprolites at Magnet Cove, Arkansas (see Chapter 4). However, all three TiO_2 polymorphs are apparently rendered less stable if they contain appreciable amounts of niobium and rare earths, as in alkalic rocks.

The stability of perovskite in the weathering zone is poorly known, in part because of its sporadic occurrence. Perovskite persists in the regolith over the alkalic stocks of Brazil as nucleii of grains heavily rimmed by anatase. In the Powderhorn district of Colorado, perovskite is locally altered to anatase even though the weathered zone is thin.

Ilmenite, considered as a fixed mineralogic entity, is residually enriched by weathering processes (Goldich, 1938; Hartman, 1959; Overstreet and others, 1963; Loughman, 1969) because it is moderately stable in the weathering environment. The ratio of ilmenite to magnetite commonly increases with progressive weathering. Over basalts, ilmenite may increase sufficiently for the saprolite to become a marginal resource (e.g., Hosterman and others, 1960).

Alteration of ilmenite

Ilmenite undergoes a remarkable transformation in the weathering environment; it is moderately stable there, but iron is progressively leached from it. The result of extreme leaching can be tan-colored grains having a specific gravity of about 3.0, a

weathering site (McLaughlin, 1955) and from one silicate mineral to the other. Sphene and melanitic or schorlomitic garnet are two titanium silicate minerals that may weather slowly and locally may survive even in saprolite and other thick soil profiles (Overstreet and others, 1963; Harben, 1984).

rutile x-ray diffraction pattern, and a TiO_2 content of over 80 percent. The economic consequences of this alteration are discussed in Chapter 1.

The progressive alteration of ilmenite in weathering is the subject of a large body of literature (Creitz and McVay, 1948; Lynd and others, 1954; Bailey and others, 1956; Karkhanavala and others, 1959; Flinter, 1959; La Roche and others, 1962; Temple, 1966; Dimanche, 1972; Grey and Reid, 1975; Dimanche and Bartholome, 1976; Subrahmanyam and others, 1982; Morad and Aldahan, 1986). Variations in microtexture, chemistry, crystallography, and magnetism of the ilmenite alteration series are well documented. The weakness of this literature is that each study is of detrital mixtures of variously altered grains in a given sand; thus the sites of weathering are not known.

Ilmenite from unweathered parent rock as supplied to the weathering environment generally contains less TiO_2 than stoichiometric ilmenite (52.6 percent TiO_2), because of intergrowths with hematite, magnetite, other oxides, and silicate inclusions. Industrial ilmenite concentrates from such rocks contain as much as 32 to 46 percent TiO_2.

Altered ilmenite, on the other hand, contains from about 54 to more than 80 percent TiO_2. Altered ilmenite that contains less than about 65 percent TiO_2 is typically black, paramagnetic, and greater than 3.3 in specific gravity. Brown to tan, nonmagnetic, less dense, porcelanous opaque material that contains more than 70 percent TiO_2 is known as leucoxene. This leucoxene contains microcrystalline rutile; leucoxene has no real mineralogic significance, however, as low-grade metamorphic leucoxene has a similar appearance but consists of sphene ($\pm$ anatase).

The trace-element content of ilmenite also varies with the intensity of weathering alteration. Based on rather fragmentary evidence, it appears that MgO contents decrease as alteration proceeds and that MnO contents increase in the first stages of alteration (to low-TiO_2 altered ilmenite) but decrease at higher TiO_2 values (Welch, 1964; Baxter, 1977, 1986; Mertie, 1979). An increase in Al_2O_3 may occur by infiltration into porous altered ilmenite (Frost and others, 1983).

Oxidation and humic acid leaching were shown by Karkhanavala and Momin (1959) and Lynd (1960) respectively, to facilitate the weathering alteration of ilmenite. Reportedly, initial alteration oxidizes iron to Fe_2O_3, and later alteration leaches iron (Temple, 1966; Garnar, 1972; Baxter, 1977).

At TiO_2 contents of about 58 to 60 percent, the ilmenite lattice is no longer detectable to x-ray diffraction, in my experience. The anisotropy and pleochroism of ilmenite in polished section also disappear (Bailey and others, 1956; Dimanche and Bartholome, 1976). Textural relics of intergrown iron oxide phases can be recognized in altered ilmenites, but at higher TiO_2 contents, these are represented only by pits. Even recognizable hematite is etched away (Temple, 1966; Puffer and Cousminer, 1982; Darby, 1984). Single grains may be concentrically zoned, from ilmenite at the core to leucoxene on the rim (Welch, 1964; Frost and others, 1983), so that the external appearance of a grain does not always correlate with composition and magnetic properties.

Recent studies appear to agree that pseudorutile (about $Fe_2Ti_3O_9$, or 60.0 percent TiO_2; Teufer and Temple, 1966) is commonly an intermediate product of ilmenite alteration. This compound may be the same as that debated in older literature under the name arizonite. Fine pseudorutile is a pseudomorph of ilmenite that retains some optical properties of ilmenite grains (Mathis and Sclar, 1980).

SITES OF TITANIUM-MINERAL WEATHERING

The weathering alteration of titanium oxide minerals takes place in four broad environments (Fig. 46). Based on the rather sketchy information currently available, it is possible to tenta-

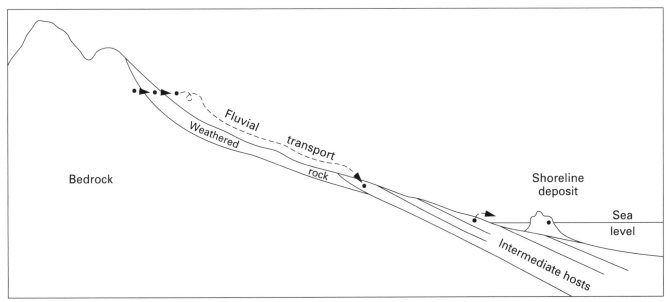

Figure 46. Diagram showing sites of weathering beneficiation of a titanium-oxide mineral grain.

E. R. Force

TABLE 15. COMPARISON OF ILMENITES FROM PAIRS OF FRESH ROCK AND SAPROLITE, IN GRANITOID ROCKS OF SOUTHEASTERN UNITED STATES*

State/Sample	Ilmenite content (%)	Ilm/mag ratio	Ilmenite (+ hematite) concentrate					
			TiO_2	MgO	MnO	Fe_2O_3/FeO	Fe oxides/TiO_2	Intergrown magnetite[†]
North Carolina								
Saprolite 1	0.71	4.4	23.08	0.01	0.97	3.89	3.29	P
Granitoid 1	0.085	5.0	19.27	0.03	0.96	4.60	4.14	P
Gain (%)[§]	-16	-12	20	-67	1	-15	-21	
South Carolina								
Saprolite 2	0.060	0.79	48.27	0.01	8.30	0.64	0.90	A
Granitoid 2	0.011	0.16	50.05	0.06	8.27	0.22	0.83	P
Gain (%)[§]	445	394	-4	-83	<1	190	8	
South Carolina								
Saprolite 3	0.053	1.32	27.34	0.01	1.41	3.95	2.61	A
Granitoid 3	0.046	0.86	26.32	0.04	1.44	3.66	2.74	P
Gain (%)[§]	15	53	4	-75	-2	8	-5	
Georgia								
Saprolite 4	0.28	>28	51.85	0.05	4.48	0.33	0.84	A
Granitoid 4	0.22	7.6	47.97	0.11	4.17	0.37	1.00	A
Gain (%)[§]	27	>268	8	-55	7	-11	-16	
Georgia								
Saprolite 5	0.037	1.76	46.18	0.01	1.75	1.06	1.13	A
Granitoid 5	0.021	1.75	42.66	0.02	1.64	0.79	1.31	A
Gain (%)[§]	76	0	8	-50	7	34	-14	
Average Gain (%)	109	>140	7	-66	3	41	-10	

*From Mertie (1979) and my own observations.

[†]Ilmenite remained constant in all pairs; hematite remained constant except in pair 4, where it decreased in saprolite. P = present; A = absent.

[§]Gain expressed as [saprolite value/fresh rock value -1] x100%.

tively partition the weathering of titanium oxide minerals among these environments.

Weathered rock

The behavior of titanium minerals in the weathered mantle over igneous and metamorphic rocks has been outlined above. Ilmenite may alter significantly in weathered rock (Jackson and Sherman, 1953; Carroll and others, 1957; Puffer and Cousminer, 1982, p. 386), but modest alteration is more common (Rumble, 1973; Dimanche and Bartholome, 1976). Ilmenite alteration in the weathering environment is locally to anatase rather than to pseudorutile and microcrystalline rutile.

Table 15 shows the ilmenite contents and compositions of five pairs of fresh bedrock and overlying saprolite from granitoid rocks of the southeastern United States. The data show some scatter, but average values show clear trends. Ilmenite is residually enriched in saprolite, commonly by 15 to 75 weight percent, and the ilmenite/magnetite ratio increases by an average of 140 percent. The ilmenite concentrate from saprolite contains slightly more TiO_2 but less MgO than that from rock. The Fe_2O_3/FeO ratio of this concentrate is higher in saprolite, but total iron in saprolite decreases relative to TiO_2. X-ray diffraction patterns of the concentrates from fresh rock and from saprolite are similar and show sharp ilmenite peaks, but ilmenite from saprolite shows subdued magnetite peaks. Hematite intergrowths with ilmenite remain unaffected. Thus, the alteration of ilmenite in these saprolites is modest. In comparison to feldspar and ferromagnesian minerals, which are almost completely decomposed in these saprolites, ilmenite alters slowly in this environment.

Fluvial transport

The second potential site for weathering of titanium oxide minerals is the fluvial transport system. Little is known about its role. Cannon (1950) asserted that significant alteration of ilmenite occurs in the fluvial transport system; however, Austin (1960) suggested that this alteration occurs during occasional burial. Riezebos (1979) found that ilmenite abundance and alteration state were little changed by transport over hundreds of kilometers in the humid tropics. Darby and Tsang (1987) and Basu and Molinari (1989) found that fluvial ilmenite clearly reflects the trace-element composition and intergrowth structure of its crystalline sources. Rivers draining crystalline rocks commonly deliver labile mineral assemblages long distances to the seacoast with little modification (Russell, 1937; van Andel, 1950; Neiheisel, 1976); ilmenite at the river mouth may remain fresh (LaRoche and others, 1962; Neiheisel, 1976; Force and others, 1982). Chapter 8 treats deposition in the fluvial environment.

Intermediate sedimentary hosts

The role of intermediate sedimentary hosts as sites of weathering of titanium oxide–mineral grains is also poorly known but is probably great. A clastic grain typically comes to rest in a succession of sedimentary deposits for varying lengths of time. At each site, the grain is subject to early diagenesis and intrastratal solution, then exposure to chemical weathering, followed by further fluvial or marine transport to the next host.

The evidence of weathering in intermediate hosts is mostly indirect; for example, the mineral assemblages in unconsolidated Upper Cretaceous through Cenozoic "coastal plain" deposits in the eastern United States vary in units of different ages (Owens, 1985), apparently as a function of paleoclimatic variation (discussed below). Compared with juvenile assemblages being supplied via fluvial systems today, mineral assemblages of such sediments are generally more mature, partly because they have been subjected to intrastratal solution (discussed in Chapter 10) if not to greater weathering. Where detritus derived from coastal plain deposits is admixed with juvenile detritus, in streams draining crystalline terranes, the fluvial mineral assemblage becomes more mature.

The mature mineral assemblages of some of these intermediate hosts include altered ilmenite, with elevated TiO_2 contents (e.g., Wilcox, 1971; Puffer and Cousminer, 1982). Detritus derived primarily from such intermediate hosts inherits their ilmenite alteration (e.g., the Horse Creek, South Carolina, placer of Williams, 1967).

Postdepositional weathering

In a given deposit, it is difficult to separate the changes due to weathering to which detrital particles were subjected separately, before deposition, from weathering changes they endured together, after deposition. The guidelines that seem most useful in separating these cases are:

1. Gradients in mineralogic composition that are related to depth within weathering or soil profiles of the deposit are attributable to postdepositional weathering.

2. Differences of mineralogic compositions that are a function of age in series of similar deposits must be due to postdepositional weathering, assuming that the nature of minerals supplied to all the deposits was constant. This topic is discussed further in the next section.

3. Detrital grains of the same original character that now show heterogeneous alteration states in one deposit must represent a detrital mixture of grains having varied alteration histories, acquired before deposition. Conversely, uniform alteration of grains of the same original character is evidence of postdepositional alteration.

4. Altered grains that show a polish acquired by abrasion in the depositional environment must have been altered before deposition, as postdepositional alteration would destroy this polish by authigenic mineral growth.

5. Chemical environments such as peat beds within a deposit may prevent postdepositional alteration. Grains preserved in these environments show only predepositional alteration.

6. The average grain size of leucoxene compared with other minerals may reflect its present density or the greater density of an ilmenite precursor (see "Depositional equivalence," Chapter 7). Where leucoxene is coarser than ilmenite, the leucoxene must have been altered before deposition (Wilcox, 1971).

The literature on placer deposits of titanium oxide minerals presents several examples of the first guideline for postdepositional alteration (e.g., Pirkle and Yoho, 1970; Welch and others, 1975). The TiO_2 contents of ilmenite concentrates, as well as leucoxene/ilmenite ratios, increase upward within soil profiles (Figs. 47, 48); these profiles may be superimposed at an angle to primary sedimentary bedding, as at Trail Ridge, Florida (Force and Garnar, 1985). Changes in the entire heavy mineral assemblage may parallel those of ilmenite alteration; the mineral assemblages are commonly most mature in the uppermost zones of weathering profiles (Force and others, 1982; McCartan and others, 1990; for dissenting views, see Hails and Hoyt, 1972; Beck, 1973).

Iron leached from ilmenite in the surficial zone may be reprecipitated deeper in the same deposit as local iron hydroxide cement (Welch, 1964; Puffer and Cousminer, 1982). This cement encases ilmenite grains and retards further leaching (Lissiman and Oxenford, 1975; Baxter, 1986). Heavy-mineral-rich laminae may be preferentially cemented by iron-rich humate (Fig. 49), or humic acid salts.

The positions of present and former water tables seem to function as a base level for some types of postdepositional alteration. These levels are also depositional loci of humate and/or iron hydroxide cements (Fig. 47). Grey and Reid (1975) proposed that some ilmenite alteration at Trail Ridge proceeds below the water table, but in Chapter 9, I explain their data in another manner. Weathering below the water table passes gradationally into intrastratal solution, a topic discussed in Chapter 10.

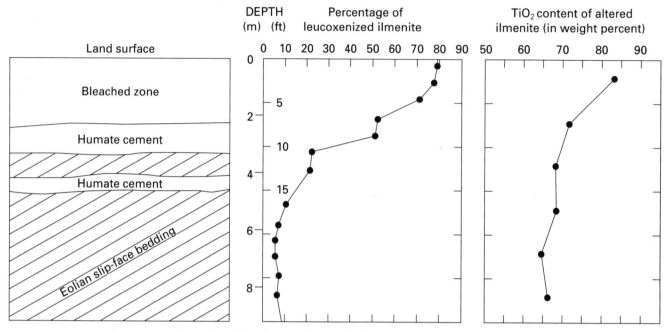

Figure 47. Vertical profiles of ilmenite alteration in surficial sands on Trail Ridge, Florida, from data in Pirkle and Yoho (1970) and Temple (1966). Geologic profile and sedimentary structures (left) after Force and Garnar (1985).

WEATHERING OF QUATERNARY SHORELINE DEPOSITS AS A FUNCTION OF AGE AND LATITUDE

A considerable literature exists on variations of heavy-mineral assemblages with the elevations of Quaternary marine terraces in the eastern United States (Neiheisel, 1962; Hails and Hoyt, 1972; Thom and others, 1972; Beck, 1973; Cazeau, 1974; Oaks and others, 1974; Force and Geraci, 1975; Force and others, 1982; McCartan and others, 1990). The higher terraces are known to be older, and each represents an interglacial high stand of sea level. The higher terraces contain the more mature, or weathered, heavy-mineral assemblages and less feldspar.

Each of these studies was done in a restricted area, and compilation of such information makes possible a display (Fig. 50) of mineralogy as a function of latitude and terrace height (the latter proxying for age). The percentage of the labile amphiboles and pyroxenes in the heavy-mineral assemblage was chosen as a measure of weathering, because these minerals show great variation in degree of postdepositional weathering from terrace to terrace. The array in Figure 50 is quite orderly; at any given latitude, weathering is greater in the older, higher terraces, whereas at any given terrace elevation, there is an equally striking increase in weathering toward lower latitudes. The small reversals in trend are due to the influx of juvenile unweathered debris at the mouths of rivers draining crystalline rock.

Similarly, the TiO$_2$ contents of altered ilmenites from the same marine terrace deposits (Force and Geraci, 1975; Force and

others, 1982) show that within a given area, ilmenite is more enriched in TiO$_2$ on the higher terraces and that within a terrace the higher values are at lower latitudes (Fig. 51). Both the mineral-assemblage and ilmenite-composition arrays show unusual simplicity and thus have predictive power. The great long-strike homogeneity of the Appalachians may have provided a relatively constant source area and thus permitted the simplicity. However, data compiled from Baxter (1977) suggest that altered ilmenite compositions on the west coast of Australia behave in the same manner (Fig. 52). Possibly, detrital mineral assemblages and ilmenite compositions are everywhere sensitive functions of age and latitude and everywhere correlate with each other. For example, correlation of Figures 50 through 52 suggests that amphibole and pyroxene become negligible when the TiO$_2$ content of altered ilmenite exceeds about 57 percent.

Compositions of altered ilmenite compiled from Quaternary shoreline deposits worldwide are shown in Figure 53. The relative ages of most of these deposits are unknown, and the nature of the ilmenite supplied by various source areas varies. Nevertheless, the data, including those of Figures 51 and 52, define a compositional envelope for altered ilmenite that is a function of latitude. At high latitude, the composition of detrital ilmenite is the composition supplied by source rocks. At lower latitude, TiO$_2$ content of ilmenite is extremely variable as a function of weathering duration and deposit type but shows a maximum at low latitude beyond which leucoxene loses coherence and density. Steep side gradients of the envelope are such that ilmenites having TiO$_2$ contents greater than about 50 percent are mostly between 35°N

TiO$_2$ content of altered ilmenite (in weight percent)

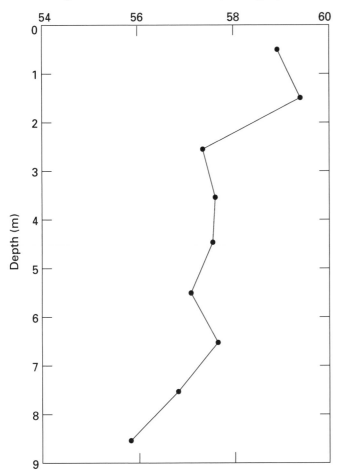

Figure 48. Vertical profile of ilmenite alteration in surficial sands of the Yoganup shoreline, western Australia. Values in Welch and others (1975) have been averaged for each depth interval.

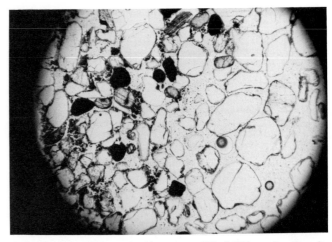

Figure 49. Photomicrograph of impregnated Trail Ridge eolian ilmenite ore sand, showing contrasts in mineralogy, grain size, and humate cementation between slip-face laminae rich (top) and poor (bottom) in heavy minerals. Heavy minerals are altered ilmenite (black), zircon, and sillimanite. Cloudy dark matrix is humate. Transmitted plane light, 2-mm field.

and 35°S. All of the economic Quaternary shoreline deposits of ilmenite occur in this zone. Thus, in deposits of this type, the distribution of altered ilmenite defines a latitude-parallel zone permissive of exploitation.

A PROPOSED MECHANISM OF ILMENITE ALTERATION

Some clues to the nature of ilmenite alteration are implied by relations described above. The unique potency of humic acids in the leaching of ilmenite implies that its oxygen-buffering capability, as well as some acidity, is required. We know from the chemical path of ilmenite alteration that oxidation is followed by some combination of acidification and/or reduction to put trivalent iron into solution. We know from postdepositional weathering profiles that ilmenite is effectively leached in a zone of shallow burial above the water table (Fig. 47). We know that humate accumulates at the water table by neutralization of humic

acids. Last, we know that humate can also precipitate in heavy-mineral–rich laminae (Fig. 49).

Accordingly, it is possible to specify a model for ilmenite weathering. Shallow burial above the water table permits intermittent saturation with surface-derived humic acid solutions. These humic acids donate oxygen for first-stage alteration but are sufficiently reduced to put iron into solution. Humic material can precipitate as humate either by consumption-neutralization on grains of altering ilmenite or by pH-neutralization at the water table. Iron put into solution by humic acid leaching can precipitate as hydroxides and/or complex with humate in the same two places.

A typical detrital concentration of titanium oxide minerals contains grains that have survived such alteration in varying numbers of intermediate hosts and to varying degrees. Each deposit records two types of alteration, a predepositional alteration of mixed degree and an overprinted postdepositional alteration limited to certain surficial zones.

CLIMATE AND PALEOCLIMATE

Preceding discussions have treated weathering as an unchanging phenomenon, but weathering is a function of climate, and climate is known to have changed through geologic time (reviewed by Frakes, 1979). Even during the short (Quaternary) period of time considered in Figures 50 through 53, important variations of climate have occurred. The strong latitude-related control implied in those figures for the alteration of ilmenite actually represents a collage of interglacial climatic conditions through Quaternary time. Ilmenite compositional gradients shown at 35° latitude may not correspond precisely with today's climate but rather with a weighted average of Quaternary interglacial climates.

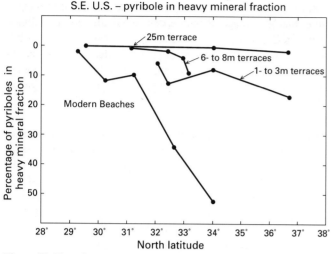

Figure 50. Plot of percent pyriboles (pyroxene plus amphibole) in Quaternary(?) shoreline deposits of the southeastern United States as a function of terrace elevation and of latitude. Data from Martens, 1935; McCaulay, 1960; Neiheisel, 1962; Thom and others, 1972; Beck, 1973; Oaks and others, 1974; and Force and others, 1982.

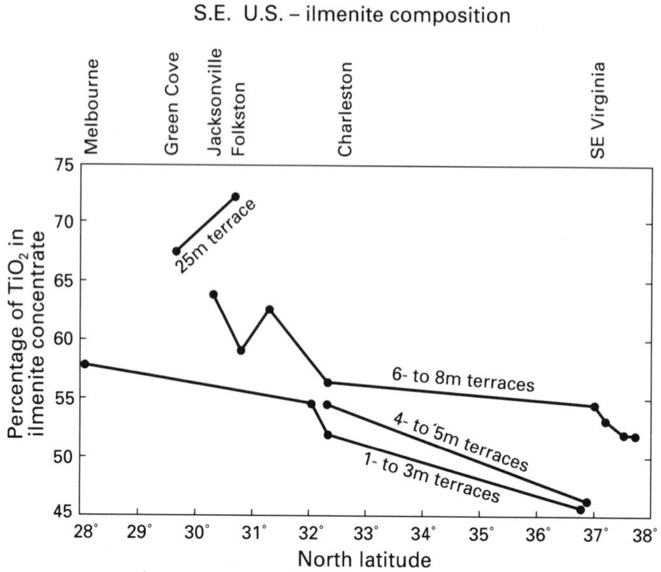

Figure 51. Plot of TiO_2 content of ilmenite concentrates from the southeastern United States, as in Figure 50. Sources include my unpublished data; unpublished data of J. B. Mertie; McCauley, 1960; Force and Geraci, 1975; Force and others, 1982; and Pirkle and others, 1984. The anomalous trend of the 25-m terrace is unexplained; the data come from different sources.

Climate change and weathering

Climatic zones on continents are basically latitude parallel, with an overprinted difference between east and west coasts. Climate change can be visualized as relative motion between a given continent and the climate belts crossing it. This motion may be of two general types. (1) Latitude-parallel belts or isotherms may move away from or toward the equator by climate change; this is potentially the more rapid motion. (2) Tectonic motion of a continent may carry it through climatic zones with fixed positions. The second factor was important in the Cenozoic for Australia and the Indian subcontinent.

Two examples of resultant climatic change that affects weathering rates are: (1) a cool moist area becomes a warm moist area, accelerating the weathering rate; and (2) a hot dry area becomes a hot moist area, also accelerating weathering.

Each geologic time period is characterized by certain climatic regimes and consequently by different areal distributions of intense weathering. These temporal variations in weathering are recorded in the detrital mineral assemblage of sedimentary rocks deposited through these time periods. The mineralogy of some of these rocks records weathering more severe than at present. Where these rocks function as intermediate sedimentary hosts for younger deposits of titanium oxide minerals, they release grains to the modern environment that are weathered beyond the power of present climatic conditions. Herein lies much of the importance of intermediate hosts in the weathering history of titanium minerals.

The Cretaceous is a good example of a time period when climate and weathering were much different than today. Temperate climates extended to high latitudes (Frakes, 1979; Arthur and

others, 1985), perhaps to 80°, and carbon dioxide and oxygen in the atmosphere were higher (Fischer, 1981; Berner and others, 1983; Berner and Landis, 1988). Deep weathering, locally to produce economic mineral deposits, occurred (Goldich, 1938; Tourtelot, 1983; Purucker, 1983). Ilmenite-bearing placer deposits that formed during Cretaceous time (Wilcox, 1971; Houston and Murphy, 1977) or were reworked from Cretaceous deposits (Horse Creek placer of Williams, 1967) commonly have mature mineral suites, including ilmenite having high TiO_2 contents. These deposits are also characterized by high ratios of monazite to total heavy minerals. The stability of monazite, a soft, dense phosphate mineral, may be sensitive to atmospheric composition. Perhaps high atmospheric CO_2, oxygen, and/or some other atmospheric gas stabilized detrital monazite during the Cretaceous.

Puffer and Cousminer (1982) have analyzed paleoclimatic controls of ilmenite weathering in a deposit of Miocene-Pliocene age in some detail. This time period was one of relatively intense weathering also. Paleoclimatic aspects of pre-Quaternary placer deposits are further discussed in Chapter 10.

Variation with Quaternary sea levels

During the Quaternary, climatic variations have been tied more or less mechanically to variations in ice volume and thus to sea levels. During peak glacial periods, sea level was about 85 m lower than today's, sea-surface temperatures averaged about 4°C cooler, and the CO_2 content of the atmosphere was about 30 percent less (Ruddiman, 1985; Oeschger, 1985; reviewed in

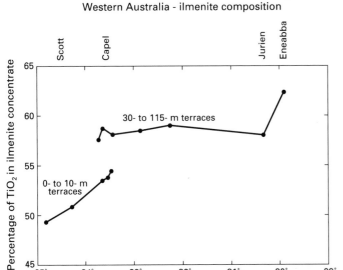

Figure 52. Plot of TiO₂ content of ilmenite concentrates in Quaternary(?) shoreline deposits of western Australia. Data from Lissiman and Oxenford (1973; Eneabba only) and Baxter (1977).

history of shoreline deposits: (1) the duration of subaerial exposure, that is, the age of the deposit minus duration of submergence; and (2) the weighted average of weathering rate while exposed, which is a function of paleoclimate. Total weathering should be a function of the product of these two factors, both of which in turn are functions of the present depth of submergence:

$$W = R\,t = f_1(D)\,f_2(D) \qquad (6\text{-}1)$$

where W is total postdepositional weathering, R is weathering rate, a function of climate, t is time intervals of subaerial exposure, and D is present depth of submergence.

Variations in weathering rate $[R = f_1(D)]$ are addressed first, and the east coast of the United States is used as an example. Coastal temperatures for each season during glacial periods have been quantified by McIntyre and others (1976). Figure 54 converts their coastal paleotemperatures to isotherms on the continental shelf; these isotherms represent summer temperatures when each point was at sea level, assuming that sea-level fall was proportional to temperature drop. Isotherm shifts of about 3° of latitude toward the south are common on the outer shelf. If these temperature variations were representative of all climate variations, the favorable zone for the formation of ilmenite placer deposits would shrink during a glacial period to latitudes lower than about 32°.

The other factor in weathering, duration of subaerial exposure $[t = f_2(D)]$, acts as a permissive condition. When a previously formed shoreline deposit is submerged below sea level, it cannot weather in the conventional sense; there is no evidence of submarine leaching of titanium oxide minerals. Thus, subaerial exposure is necessary for weathering of mineral suites of interest. A shoreline deposit that formed at sea level but is now at 60-m depth on the ocean floor would seldom have been exposed as sea level cycled through the range of glacially controlled stands.

Sundquist and Broecker, 1985). The smoothly cyclical and repetitive nature of both the sea-level and paleoclimatic Quaternary cycles suggests that certain Quaternary sea levels represent certain climates, independent of time. If so, we can address the subaerial weathering history of Quaternary shoreline deposits formed at various sea levels, including those now offshore. This approach is invalid for tectonically active coasts, including those rebounding from ice load.

Two factors interact to determine the subaerial weathering

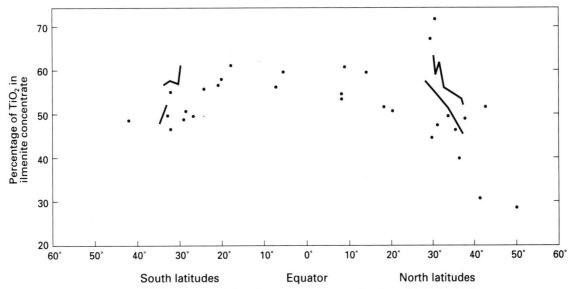

Figure 53. Plot of TiO₂ content of ilmenite concentrates as a function of latitude, in Quaternary shoreline deposits of the world, compiled from numerous sources, published and unpublished. Data from Figures 51 and 52 are summarized as lines.

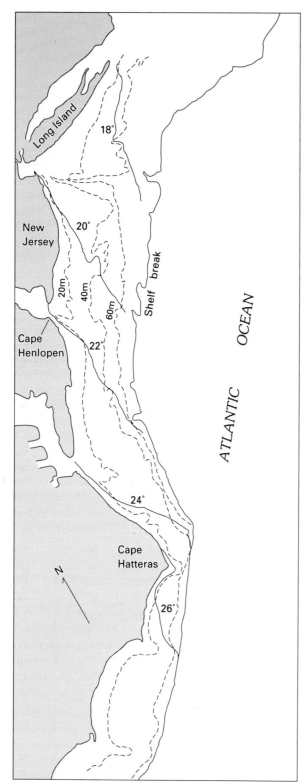

Figure 54. Map of the continental shelf off the eastern United States showing summer isotherms in degrees centigrade (solid lines) applicable when each point was at sea level. Bathymetric contours (dashed) from National Oceanographic and Atmospheric Administration. Quaternary paleotemperatures from McIntyre and others (1976).

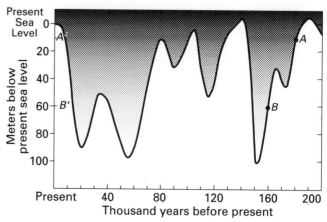

Figure 55. Quaternary sea-level curve with shading proportional to weathering intensity at each sea-level position. Weathering history of deposits A and B contrasted in text.

The two factors act in concert in such a way that Quaternary shoreline deposits found at progressively greater depths offshore are progressively less weathered. A shoreline deposit now at great depth has seldom been exposed, and when it was, the temperature was such that it did not weather rapidly. Observed variations in the mineralogy of offshore sands (Reich and others, 1982) suggest that immature assemblages are found at greater depths. This principle could apply even to shoreline deposits that are smeared out by later current reworking.

Figure 55 is a diagrammatic solution of the interaction of the weathering rate and duration variables (Equation 6-1). It is a sea-level curve on which are superimposed weathering-intensity variations that correspond to each sea-level position. Points A and B represent two shoreline deposits of about the same age that were formed during different phases of the same sea-level cycle. Deposit A received weathering proportional to all shading along the sea-level curve below line A-A'. The deposit at B receives much less weathering, proportional to shading below line B-B', probably about 25 percent as much as A.

EXPLORATION IMPLICATIONS

Weathering variations have exploration implications for weathering-beneficiated deposits of titanium oxide minerals in igneous and metamorphic deposits and for all placer deposits. In deposits of Cretaceous age, weathering beneficiation probably occurred on all stable, moderate-relief land surfaces at latitudes up to at least 50°, except where the landmass more recently moved into such latitudes from higher latitudes. For younger deposits this beneficiation situation is more constricted. For Quaternary deposits formed during interglacial periods, the favorable latitudes of syn- and postdepositional weathering are lower than 35°. For deposits formed at the lower sea levels of glacial periods, favorable latitudes may not exist.

Chapter 7.

Titanium oxide minerals in sedimentary rocks and principles of placer concentration

DISTRIBUTION AMONG SEDIMENTARY ROCKS

The chemical immobility of titanium has produced a predictable distribution pattern of titanium in sedimentary rocks. Chemical sediments are systematically impoverished. Detrital sediments have titanium contents controlled by the mechanical behavior of titanium-bearing grains in various hydraulic regimes. These grains may be either titanium-bearing silicates or oxides.

Limestones average 0.1 to 0.2 percent TiO_2 (Table 1). Even in these chemical sediments, TiO_2 is commonly present as detrital admixtures, as in eolian sand fractions containing ilmenite.

Shales average 0.6 to 0.7 percent TiO_2 (Table 1), and values over 1 percent are common; shale is the sedimentary rock with the highest average TiO_2 content. Because of fine grain sizes, the titanium mineralogy of shale has never been adequately studied. I suspect that much of the TiO_2 in shales represents the fine-grained anatase present in regoliths as a result of the weathering of titanium-bearing silicates. Siltstones have slightly lower average TiO_2 contents than do true shales.

Sandstones and coarser clastics vary greatly in TiO_2 contents because of variations in concentration of titanium-bearing grains; the average value must be 0.25 to 0.5 percent TiO_2. The TiO_2 also shows great variation in partitioning between oxide and silicate minerals. In the mineralogically immature sandstones, titanium mineralogy closely follows that of source rocks. This typically means that much titanium is present as silicate grains, particularly in the active continental-margin settings where most of these sands are formed. For example, in voluminous terranes of volcanogenic graywackes, representing many depositional environments, sandstones contain titanium as magnetite, ilmenite, titanaugite, hornblende, and/or sphene. This suite of titanium minerals may occur in placer concentrations of heavy minerals, resulting in deposits having high TiO_2 contents but no economic value (e.g., Thomas and Berryhill, 1962).

Mineralogically mature sandstones typically contain TiO_2 mostly as altered ilmenite and rutile, as by definition all labile minerals, including titanium silicates and magnetite, are absent. The restricted mineralogy of these sands is acquired by multiple cycles of chemical and mechanical beneficiation, commonly in passive-margin or cratonic tectonic settings. Because only a few minerals can contain titanium in these sands, titanium enrichments in mature sands are unusual. However, mature sands in which placer concentration has occurred have moderate to high TiO_2 contents as rutile and ilmenite and are economically valuable deposits. The remainder of this book is devoted to discussion of these deposits and the processes forming them.

PLACER CONCENTRATION

A placer-enriched deposit is formed by the concentration of denser minerals. This concentration occurs by a complex interaction among transport type, particle size and shape, and particle density. I shall attempt to explain this interaction in nonmathematical terms. For any type of transport or erosion, a given flow removes some particles from the bed and leaves others. The division is primarily a function of particle size and density. For the given flow, a particle of quartz or feldspar of a certain grain size will behave like a denser mineral of a different grain size. The laws relating particle size and density vary among different transport modes, and this variation is the key to placer concentration; adjustment by the bed to successive types of transport and erosion can enrich dense minerals.

Transport laws vary at a number of scales; for example, the laws for a beach deposit and for the adjacent eolian dune differ. At progressively smaller scales, within the beach deposit there are different laws for breaking-wave and backwash transport and even for different portions of a single migrating ripple.

The laws governing equilibrium transport as a function of density and grain size have been studied for many modes of transport and erosion. The law studied first (Rubey, 1933) is probably the simplest—deposition from suspension, such that grains of the same settling velocity accumulate together. Settling velocity is, of course, a function of density and grain size. The term *hydraulic equivalence* was preempted for this type of equilibrium (Rittenhouse, 1943), but clearly a whole family of hydraulic equivalences exists for different transport types. These can be loosely divided into depositional equivalences (which include suspension equivalence) and entrainment equivalences, with transport equivalence a complex mixture of the two (Slingerland, 1984; Slingerland and Smith, 1986). In some deposits, dispersive equivalence (natural heavy-media separation) is also a factor. Each equivalence type is described below.

The enrichment of dense minerals (relative to light minerals) is not directly explained by any one law of hydraulic equivalence, because in deposits formed according to that law, heavy minerals will be deposited with hydraulically equivalent light minerals, and no concentration is achieved (Force, 1976b; Komar and Wang, 1984; Slingerland and Smith, 1986). In only one circumstance can an enrichment result from deposition obeying one law—the case in which the dense mineral of interest is supplied in a range of grain sizes equivalent to a size range for light minerals that is in short supply. Except perhaps for minerals having specific gravities more than 8, that is, minerals hydraulically equivalent to much coarser light minerals (Tourtelot, 1968), such enrichments are rare.

The general case for placer concentration is the sequential operation of two slightly different laws in such a way that the sediment can obey both laws only by becoming enriched in small dense minerals. Most commonly, one law is of the deposition type and the other is of the entrainment type (cf., Komar and Wang, 1984). For example, deposition from a turbulent breaking wave on a beach face is essentially by a law of suspension equivalence, with any given spot in the swash zone representing a certain settling rate. Large light grains are deposited with small dense grains (Fig. 56, top). The more laminar backwash of the same wave, however, has an erosive effect since its suspended load has been dropped. It erodes according to an entrainment equivalence law that is almost entirely a function of grain size; the grains that project out of the bed are eroded, and these are mostly coarse (Slingerland and Smith, 1986). Thus the large light grains selected by the first process are removed by the second, and an enrichment in small dense grains results (Fig. 56, bottom). The enriched deposit remains in equilibrium with the next wave. Note that if the first process had been inefficient, large dense grains and fine light grains would also have been deposited. The latter would not have been eroded by the second process, and little or no enrichment would have occurred (Fig. 56, center). This simplified example is expanded in Chapter 9.

Depositional equivalence

Because of difficulties of direct observation, laws of depositional equivalence have mostly been studied experimentally and theoretically. The difficulty of observation in a natural situation

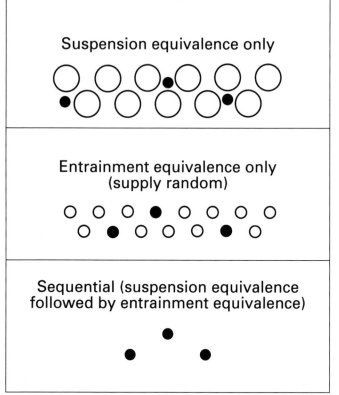

Figure 56. Diagram of grain populations of light (open) and heavy (dark) minerals resulting from various hydraulic equivalence laws.

comes in the separation of depositional and entrainment components of transport. In our wave example, thorough sampling between breaking wave and backwash is difficult (however, see Slingerland, 1984).

Rubey (1933) formulated the concept that grains found together forming a single bed represent grains having equal settling velocity, that is, coarser light minerals and heavy minerals finer in proportion to their densities. Rittenhouse (1943) called this condition hydraulic equivalence and cast his similar results with fluvial sands in the form of hydraulic ratios relating the diameters of light and various heavy grains deposited together. Ironically, though the more sophisticated treatments of heavy mineral deposition have moved significantly beyond these landmark papers, most studies of heavy mineral assemblages are not up to the conceptual level of Rubey and Rittenhouse (reviewed by Force and Stone, 1990).

Figure 57 illustrates the working of depositional equivalence in a single bed. This example is from foreset beds of a glaciolacustrine delta, and deposition was from suspension in continuous turbidity flows (Force and Stone, 1990). The size distributions of various minerals supplied to the depositional environment are shown in the upper part of the figure. The size distribution of light mineral grains deposited from a given flow, to form a given bed, are shown below in solid lines. Cumulative curves for some heavy mineral species are shown superimposed on this size distribution; the position of the curve for each mineral is a function of density and shape of the mineral. The relation of its size distribution to the size distribution of the whole sample is fixed; that is, the family of curves in the bottom part of the figure slides in its entirety to the right or left as modal grain size of the bed changes as a function of flow parameters. An arbitrary distribution representing an adjacent bed is shown with dashed lines. The abundance of a given mineral in each bed depends on the relative position of the size of that mineral required for deposition from the flow (bottom of figure) and the size of the mineral in abundant supply (top of figure). Where the two coincide, the mineral will be relatively abundant. In the example shown, ilmenite and biotite will be abundant in the bed represented by the solid line, whereas garnet will predominate in the bed represented by dashed lines. Such glaciolacustrine deposits are further discussed in Chapter 8; in the present discussion, these and analogous depositional-equivalent deposits are protores for further enrichment by entrainment and transport, which have other equivalence laws.

Entrainment equivalence

Most recent studies of heavy mineral deposition have emphasized entrainment equivalence (McIntyre, 1959; Hand, 1967;

White and Williams, 1967; Grigg and Rathbun, 1969; Lowright and others, 1972; Slingerland, 1977, 1984; Komar and Wang, 1984; Slingerland and Smith, 1986). In general, these studies have shown that entrainment is a function more of grain size than of grain density. Large grains may project out of the bed into a zone of turbulent flow and thus be preferentially subjected to removal. In addition, larger grains can be preferentially dislodged by rotation through smaller angles. In a beach example studied by Komar and Wang (1984), larger grains of all mineral species are removed from the upper swash zone and transported progressively offshore. As the entire beach is fed coarser lights and finer heavies, entrainment results in a placer enrichment in the upper swash zone, which gives way seaward across the beach face to coarser-grained, less enriched deposits showing a spectrum of dominant heavies having progressively lesser density.

Transport equivalence

Fine grains commonly lag behind coarser grains in transport over a coarse bed as a consequence of entrainment differences. Where heavy minerals are finer than depositionally equivalent light minerals, this results in dynamic lag enrichments of heavy minerals (Slingerland, 1984). Where heavy and light minerals are supplied to the transport system at the same average size, heavy minerals lag because their greater settling velocities cause them to strike the bed more frequently. Where the depositional system is arrested, the lesser transport rates of denser minerals result in local concentrations.

Dispersive equivalence

In dense grain suspensions, grain collisions and fluid pressures may permit natural heavy-media separations of light and heavy minerals. Such separation depends not on flow parameters but on factors that permit the bed to remain liquified. Dense minerals sink to the base of the liquified layer. On beaches, for example, this may occur by wave-induced lateral gradients in the local water table and/or by pounding of the surf. Sallenger (1979) has shown that differential settling of denser and/or finer grains through the bed may result in both basal placer enrichment and inverse grading of beach deposits.

ECONOMIC PLACER DEPOSITS OF TITANIUM MINERALS—A SYNTHESIS THUS FAR

It is now time to gather some threads leading from previous parts of this book. These threads are some necessary preconditions for the formation of economic placer deposits of titanium minerals.

Source rocks must supply an appropriate mineral suite. The most favorable source terranes for detrital ilmenite and rutile are high-grade metamorphic rocks, locally with related igneous rocks of the charnockite-ferrodiorite-anorthosite suite (Chapters 2 and 3). These terranes are commonly exposed over areas of thousands

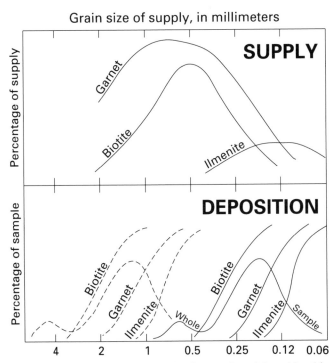

Figure 57. Grain size distributions showing supply and depositional interactions in depositional equivalence. Supply curves shown above; size distributions of two beds, and minerals thereof, shown below (one bed solid, other bed dashed). All curves are frequency diagrams except for the nested cumulative curves for individual minerals in the beds shown below. The interaction between supply and depositional curves is discussed in the text.

of square kilometers, and ilmenite and rutile dominate their titanium-mineral assemblages. Rutile is most common in pelitic lithologies (Figs. 2, 3). Erosion of these terranes releases about 0.1 to 1.0 percent ilmenite and rutile to the transport system (reviewed for rutile by Force, 1980b).

Other large source terranes supply the wrong mineral suites; for example, basalt terranes supply ilmenite intergrown with magnetite; some granitic terranes supply a little ilmenite, intergrown with hematite and magnetite; and neither terrane supplies rutile. The only other source terranes that contribute valuable assemblages of titanium minerals are small—for example, alkalic stocks and hydrothermal systems (Chapters 4 and 5). Locally, intermediate sedimentary hosts are the most important sources, but these in turn have primary source rocks. If the source terranes do not supply the right minerals, an economic titanium-mineral placer cannot form.

Weathering, as predepositional weathering of constituent grains and/or as postdepositional weathering of a placer deposit, is necessary also (Chapter 6). It restricts the mineral suite to one in which titanium minerals and other economic heavies are among the few heavy species, and it chemically upgrades the ilmenite. Currently, no titanium-mineral placers are operating in which this weathering upgrading has not occurred.

Discussion in upcoming chapters treats further controls on placer location and formation. Conduits, in both fluvial and shoreline environments, are necessary to bring favorable detritus to the depositional site in undiluted form. Once placer concentration has occurred, it must be preserved; locally this has required removal of the concentrated sediment to another depositional environment.

Thus, consideration of placer concentration processes, as in this chapter, requires consideration also of the numerous other influences on placer formation (Fig. 58A). Most of these influences are not apparent at the site of concentration. A great deal of the geologic evolution of an economic placer deposit is hidden from view at the depositional site and must be studied elsewhere.

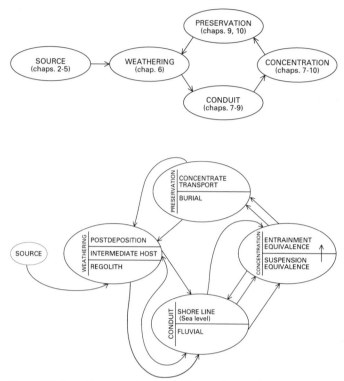

Figure 58. Box diagrams of sequential factors involved in formation of economic placer deposits of titanium minerals. A. General case. Each box has the potential of containing economic deposits. B. Adaptation of general case for Quaternary shoreline placer deposits.

Even concentration itself is a two-stage process, and the first stage may occur in part outside the depositional environment. Inherited characteristics of the placer extend from the depositional site all the way back through the geology of titanium oxides, and these inherited characteristics control the economics of exploitation just as surely as does final placer concentration.

Chapter 8.

Placer deposits of titanium oxide minerals in nonmarine sediments

Althouth numerous nonmarine placer deposits of titanium minerals have been identified, only one, a fluvial deposit in Sierra Leone, is currently of economic importance. Nature operates under too many disadvantages in nonmarine environments for there to be many economic nonmarine placer deposits of titanium minerals. Deposition in these environments is of interest partly as a step in the formation of shoreline placer deposits.

This chapter concentrates on nonmarine placers in young hosts, because virtually all identified nonmarine titanium-mineral deposits formed during the present geomorphic cycle (Table 16). The significance of youth in these deposits is apparently that they are unindurated, and they were easy to find using present geomorphology. However, older fluvial deposits, some of them indurated, were important intermediate hosts of titanium oxide minerals in economic placers of younger shoreline deposits.

RELATIVE IMPORTANCE OF GEOLOGIC CONTROLS

All nonmarine placer deposits of titanium oxide minerals appear to be intimately related to favorable source rocks (Table 16). The source factor is more important than the other factors that normally enter into the formation of a titanium oxide placer deposit (Fig. 58A), because the other factors are feebly developed. Weathering beneficiation varies greatly but is far less potent than in some shoreline deposits. True placer concentration of titanium minerals in nonmarine deposits is apparently effective only at scales much smaller than that of a mineable deposit. No large deposits formed in nonmarine environments seem to have had titanium minerals greatly enriched by placer concentration compared with the debris fed from the source.

Because of the importance of the source terranes in the formation of nonmarine placer deposits of titanium minerals, valuable deposits represent unusual geomorphologies and clastic distribution systems that minimize dilution by debris from unfavorable sources. The relative importance of nonsource factors differs in the two major types of nonmarine deposits; fluvial deposits are more beneficiated by weathering, whereas glaciolacustrine deposits are better winnowed but completely unweathered.

FLUVIAL DEPOSITS

Heavy minerals form concentrations in a large number of fluvial environments, and such concentration of different heavy-mineral species in different environments is the subject of a voluminous literature (abstracted by Smith and Minter, 1980; Slingerland and Smith, 1986). Scales of placer enrichment vary from individual bed forms to facies of large depositional systems. Much of the information available to us concerns modern enrichments (some from experiments) at the intermediate scale of point bars and channel junctions (Adams and others, 1978; Schumm and others, 1987). Larger-scale enrichments apparently form by selective preservation in particular environments of smaller-scale enrichments. The largest scales of fluvial placer concentrations, in wet alluvial fan environments, are mostly of minerals having specific gravities more than 6.

Concentration and weathering limitations

Placer concentration in fluvial environments has formed important economic deposits of a number of heavy mineral commodities, such as gold, diamonds, monazite, and cassiterite. Thus it is strange that titanium oxide minerals are extensively produced from placer deposits but rarely from fluvial ones. There are two reasons. First, the scale of fluvial placer concentration is too small. Economic placer deposits of titanium oxide minerals normally contain at least a million tons of mineral concentrate. Few fluvial bodies of this size contain concentrations of minerals having specific gravities of 4 to 5, such as rutile and ilmenite.

TABLE 16. NONMARINE PLACER DEPOSITS OF TITANIUM OXIDE MINERALS*

Deposit or district	Location	Age	Titanium Mineral	Source Terrane	Distribution System[†]	References
Fluvial						
Gbangbama (Sherbro)	Sierra Leone	Quaternary	Rutile	Kasila granulite-facies gneisses	1	Raufuss, 1973
Sand Canyon, Pacoima Canyon	California	Quaternary	Ilmenite	San Gabriel Mountain ferrodiorites[§]	2, 3	Oakeschott, 1958; Industrial Minerals, 1986
Otter Creek	Oklahoma	Quaternary	Ilmenite	Wichita Mountain gabbros, etc.	2	Hahn and Fine, 1960
Shooting Creek	North Carolina	Quaternary	Rutile	Garnet-mica schist	2	Hartley, 1971
Horse Creek	South Carolina	Quaternary	Ilmenite, rutile	Cretaceous sands	2b	Williams, 1967
Roseland	Virginia	Quaternary	Ilmenite, rutile	Roseland crystalline rocks[§]	2	Minard and others, 1976; Herz and Force, 1987
Glaciolacustrine						
Port Leyden	New York	Pleistocene	Ilmenite	Adirondack Mountain charnockites, gabbros, and granulite-facies gneisses[§]		Stone and Force, 1980

*Not including the U.S.S.R., China, or operations recovering titanium minerals as byproducts.
[†]See text and Figure 59.
[§]Chapter 3.

Second, fluvial deposits contain ilmenite of roughly the same composition as that supplied by source rocks. Shoreline placer deposits have the advantage of greater ilmenite alteration. In the only major economic fluvial deposit, the Gbangbama district of Sierra Leone, the detrital mineral of value is rutile, which needs no chemical beneficiation by weathering.

Table 16 lists the better-documented economic and near-economic fluvial deposits of titanium minerals. Titanium-mineral contents in these deposits are only slightly greater than in source rocks. Most beds in these deposits are poorly sorted, and consequently there is not much concentration.

Ilmenite in these deposits is relatively unaltered, but weathering beneficiation of fluvial deposits does have a considerable effect in restricting heavy mineral suites toward an economic assemblage. All the fluvial deposits listed in Table 16 are in areas of deep weathering. Even the San Gabriel, California, deposits, which formed adjacent to an area of high relief, show some modification of mineral assemblages by weathering. The economic role of disaggregation and weathering probably exceeds that of placer concentration in most fluvial deposits of titanium oxide minerals.

Geomorphic and lithologic controls

Three distribution systems (Fig. 59) that permit little dilution favor the formation of fluvial deposits of titanium minerals: (1) Radial drainages from positive favorable source areas. This

situation in the Gbangbama district has produced the most valuable single deposit in nonmarine deposits. (2) Drainage basins entirely within favorable source rocks. A subtype is along-strike drainages with few tributaries in elongate favorable terrane (type 2b, Fig. 59). (3) Drainage basins having headwaters in favorable source rocks and surrounded by other sediments derived from those favorable sources. The third type of system can be coupled with either of the first two, thereby extending them downstream. Most known fluvial titanium-mineral deposits are of one or more of these types (Table 16).

Within favorable drainage basins, titanium oxide minerals are distributed in predictable ways. In the fluvial deposits that have been studied sufficiently, titanium oxide minerals decrease away from the source (Hahn and Fine, 1960; Raufuss, 1973; Minard and others, 1976). This may be in spite of an increase in the proportion of titanium oxides in the heavy fraction by attrition of less stable heavies (Raufuss, 1973). The trend is due to some combination of lag effects and dilution by other debris.

Lithologically, the titanium-oxide mineral deposits listed in Table 16 occur mostly in gravelly silty sands. For example, in alluvium of the Roseland district of Virginia, the 37 layers with greater than 3 percent heavy minerals, out of 122 layers reported by Minard and others (1976), average 11.1 percent gravel (>2 mm) and 25.5 percent silt and clay (<0.06 mm). Trask sorting values (So) range from about 2 to 6, that is, the deposits are generally poorly sorted. Concentrations greater than 10 percent of heavy minerals occur in deposits that are better sorted, including some that are muddy or gravelly but not both. Heavy mineral mode was 0 to 1 ϕ finer than the mode of the entire sample (Fig. 60A) and varied from about 1 to 3 ϕ (0.125 to 0.5 mm). That is, heavy minerals are the transport equivalents of coarse to fine quartz sand in these deposits.

Commonly, the heavy-mineral content increases with depth within the alluvial sequence (Hahn and Fine, 1960; Raufuss, 1973; Minard and others, 1976). Probably the basal deposits are less muddy and represent axial-channel environments rather than point-bar or flood-plain deposits.

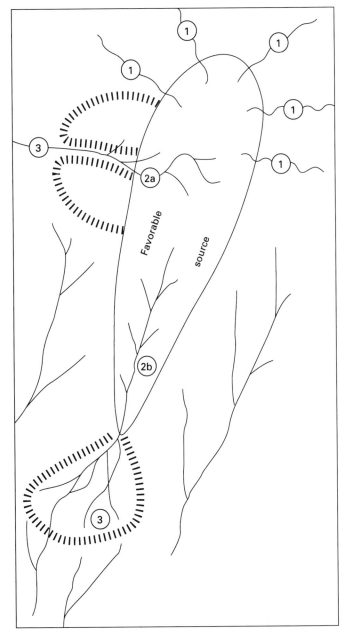

Figure 59. Diagram of fluvial distribution systems (numbered) that permit transport with little contamination by mineral suites other than those from favorable sources (discussed in text). Drainage type 1, radial; 2, within source; 3, within debris of favorable source.

GLACIOLACUSTRINE DEPOSITS

During the Pleistocene, much glacial debris was sorted and deposited in Gilbert-type tripartite deltas in numerous large glacial lakes. In deltas derived from crystalline rocks, Force and others (1976), Stone and Force (1980, 1983), and Force and Stone (1990) have found that entire foreset sequences, containing up to 10^8 metric tons of sand, are heavy-mineral deposits containing 4 to more than 10 percent heavy minerals. In the sense of Chapter 7, these show only the first stage of placer enrichment.

The foreset sequences, as thick as 80 m or more, are largely sandy and consist of beds in unconformity-bounded sets, dipping 10° to 40° toward the lake basin. Unconformably overlying the foreset sequence are cobbly fluvial topset beds (Fig. 61A).

Individual foreset beds range from coarse to very fine sand. Gravel beds and pebbly sand beds are also common. Sedimentary structures are planar lamination and climbing-ripple cross-lamination, the latter more common in finer beds. Sorting is quite good (So 1.3 to 1.6) in most foreset beds, including virtually all sand beds and some gravel beds, but is poor in bimodal gravelly sands. Sorting values for monomineralic populations are commonly as good as So 1.2. Titanium oxides are enriched in size fractions about 1 ϕ grade finer than that of the sand-sized mode for the host bed (Fig. 60B). Heavy-mineral contents are lower in gravelly beds, because of dilution of the sand fraction. Variations with stratigraphic positions or sedimentary structure were not detected in the Connecticut deltas studied by Force and Stone (1990).

The deposits are derived from specific local source terranes via glacial action, and are transported in meltwater and deposited on delta fronts too rapidly for any significant weathering to occur. Thus, in these deposits, source rocks are of great importance; Force and Stone (1990) show that adjacent coeval deltas with different sources have correspondingly different detrital mineralogies. Apparently little mixing of debris occurs in the subglacial environment, and the meltwater dispersal path is short.

Deposition of the foreset beds is by bottom-hugging turbid currents that transport the entire fluvial sediment load down the steep foreset slope (Gustavson and others, 1975). These currents are continuous but variable and shift laterally. Deposition is from suspension but with a variable traction component. Since the currents are continuous and since deposition is directly from the current, a given current deposits a bed with a narrow range of settling velocities; measured S_o values imply settling velocities differing by ±45 percent for the central half of the grain size distribution. Sorting this good is unusual for deposition from suspension, because most suspension deposition is from rapidly waning currents or from sediment rain through still water (as in marine deltas); under these conditions, grains of different settling velocities are deposited together.

In the prodelta environment, erosion probably occurs as a consequence of lateral shifting of currents, thus producing unconformities between sets. Within sets, the sedimentary structures, especially steeply climbing ripples, show that reentrainment of

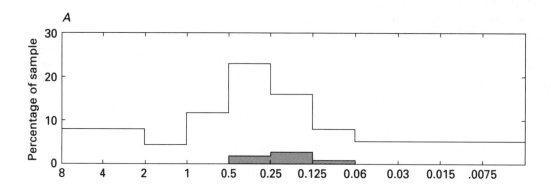

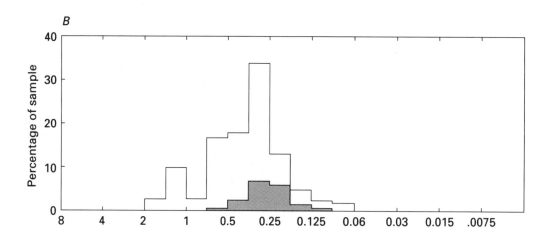

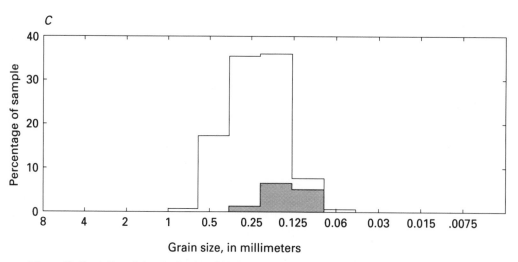

Figure 60. Examples of size distributions of light and heavy (shaded) minerals in some placer-forming environments. A. Fluvial deposit, Roseland district, Virginia; sample 328 of Minard and others (1976). B. Glaciolacustrine deposit, Connecticut; sample K12 of Force and Stone (1990). C. Eolian heavy mineral enrichment, Trail Ridge, Florida; Force and Rich (1989).

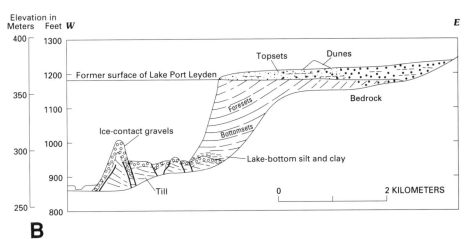

Figure 61. Structure of typical glaciolacustrine deltas. A. Topset and upper foreset beds, Hopeville delta, Quinebaug valley, Connecticut. Exposure about 5.5 m high. B. Cross section of the Port Leyden delta, New York (7.2 km across).

previously deposited grains is minor. Thus placer concentration is arrested at the first stage, the establishment of depositional (suspension) equivalence.

In glaciolacustrine foreset beds, the grain size of a bed exerts strong control over the mineral assemblage in the bed. Figure 57 illustrates the relation of heavy-mineral deposition to grain size of host and to the nature of bulk feed from source in these deposits (see text of Chapter 7 for full explanation). Given the nature of this bulk feed and the modal grain size of each bed, the mineralogy of the bed can be predicted rather precisely. For example, coarser beds may be dominated by garnet and finer beds by titanium oxide minerals. Actual enrichment in titanium oxide minerals over that in source rocks is uncommon in this environment; it occurs only where the grain sizes of titanium minerals and of hydraulically equivalent light minerals are fortuitously those in which the ratio of oxide minerals to light minerals is high in the bulk feed from source areas.

MAJOR DEPOSITS

Gbangbama district, Sierra Leone

Rutile has been mined from the Gbangbama district in the coastal region of Sierra Leone since 1967. Economically, these ventures have had a spotted history, but currently, under Sierra Rutile, the operations are a major rutile supplier. Sierra Leone is surely the only country for which rutile is the most important earner of export revenue.

The district consists of the Gbangbama Hills, about 200 m high and trending northwest to southeast, surrounded by low coastal plain. The vegetation is rain forest, and the area was remote until the coming of mining operations (Spencer, 1964).

My description of the deposits is largely from Raufuss (1973). Additional information is from Spencer and Williams (1967) and Lang (1970).

Rocks of the Gbangbama Hills are deeply weathered garnet amphibolites and leucocratic garnet granulites (Raufuss, 1973). The latter characteristically contain rutile in concentrations of about 0.2 to more than 1 percent. Besides garnet and rutile, the granulites contain two pyroxenes, ilmenite, apatite, zircon, and locally sphene. The garnet amphibolites also contain local rutile, but as these rocks are more resistant to weathering, this lithology is less well represented in derived alluvial sands.

Rutile-bearing deposits are radially disposed around the Gbangbama Hills. The present drainage does not reflect this radial pattern. At least some of the deposits are late Pleistocene in age, based on ^{14}C of lignitic layers. Characteristically, they are poorly sorted sands, commonly gravelly and averaging 35 to 45 percent muddy matrix, locally with clay layers. The deposits are mostly about 10 m and up to 20 m thick, lying directly on weathered bedrock. Many of them form discrete inliers in crystalline terrane.

The deposits themselves contain as many as two laterization surfaces. Differentiation of laterized sediment and saprolite is locally difficult (Raufuss, 1973), and this difficulty is said to have hampered earlier mining. Multiple weathering cycles clearly have affected this deposit, and the extent of mineralogic beneficiation is significant.

The detrital mineral assemblage of the deposit is actually dominated by rutile. Ilmenite contents are roughly inverse to rutile contents (probably because this ratio varies in the source terrane). Corroded pyropic garnet is a locally major constituent; it is intergrown with rutile. Other locally common minerals are amphiboles, both pyroxenes, kyanite, sillimanite, zircon, and graphite. Minor constituents include magnetite, monazite, sphene, corundum, and tourmaline. Iron oxides record the former presence of mineral grains unstable in the weathering environment. Total heavy-mineral contents commonly range from 1 to 5 percent, with rutile contents ranging from 0.5 to 2 percent.

Raufuss (1973) ascribes deposition of the most poorly sorted deposits (up to 60 percent silt and clay) to sheet floods and remarks on the ineffectiveness of true placer concentration here. Rutile content shows a high at an optimum transport distance of a few kilometers, apparently as a result of the interaction of favorable and unfavorable transport factors. Immediately adjacent to the source is a zone of poorly sorted gravel and slumped laterite; outside this is a zone of some hydraulic sorting. In this second zone, heavy-mineral concentration decreases away from the source because of decreasing grain size and possibly lag effects. The rutile portion of the heavy-mineral fraction increases in this zone, however, because of progressive elimination by weathering of garnet, amphibole, pyroxene, and magnetite. Raufuss notes that deeper deposits having coarser grain sizes commonly have higher rutile contents.

Mining thus far has been in the area northeast of the Gbangbama Hills near Mogbwemo. Deposits on the southwestern side are similar and perhaps equally extensive, but they are slightly lower in grade and show some marine influence.

The unusual mineral assemblage of the Gbangbama district is related to an unusual source, little dilution, and weathering beneficiation. Placer concentration played a minor role in the formation of economic deposits here.

Similar occurrences of rutile-dominated detrital deposits are known to extend toward the northwest, in similar relation to Kasila granulites (Raufuss, 1973). To the southeast these rocks extend across the Mano River into Liberia, where I have studied them near Lake Piso. Here they consist of equigranular leucocratic banded quartz-feldspar-garnet gneisses with minor pyroxene and 0.25 to 1 percent rutile. Interlayered mafic granulite gneisses (gnm of Thorman, 1977) containing ilmenite locally predominate. The rocks are deeply weathered, and some of the fluvial drainages allow little dilution. Thus, the Liberian end of the province has some potential for additional fluvial rutile deposits.

Port Leyden delta, New York

Along the western margin of the Black River valley, forming the southwestern margin of the Adirondack Mountains, plateaus whose flat tops are at 1,200 ft (370 m) rise as much as 80 m above the valley floor (Fig. 61B). The plateaus represent coalesced Pleistocene deltas deposited in glacial Lake Port Leyden (Fairchild, 1912), deposited from meltwater of the receding continental glacier and derived from the Adirondacks. The deltas consist largely of sandy foreset and proximal bottomset beds averaging 3.5 percent heavy minerals (Force and others, 1976; Stone and Force, 1980). The sands are well to moderately sorted and contain sedimentary structures indicative of deposition from bottom currents; among these are ripple cross-laminae with variable angles of climb, in which heavy minerals are enriched.

Ilmenite resources of about 26 million metric tons are contained in these deltaic beds in the Port Leyden Quadrangle (Force and others, 1976; mapped by Howard University Geology Field

Camp, 1974). Areas to the north and south apparently contain substantial additional resources. Ilmenite-bearing grains average 35 percent of the heavy mineral fraction; these contain locked silicate impurities and are completely unweathered. The ilmenite is approximately stoichiometric in composition, but ilmenite-bearing grains as a concentrate contain only about 25 percent TiO_2. Other minerals of value in these sands include zircon, sillimanite, and minor rutile. Pyroxene, amphibole, and garnet, however, are the most abundant constituents of the heavy-mineral fractions. Feldspar is abundant in the light fractions. All the grains are poorly rounded.

Titanium oxide minerals as byproducts from fluvial deposits

Fluvial deposits of value for other commodities locally contain titanium oxide minerals recoverable as a byproduct; in only two examples is recovery of titanium minerals occurring at present. The first is ilmenite from alluvial tin deposits of Malaya, Indonesia, and Thailand (Flinter, 1959; Macdonald, 1971a, b; Industrial Minerals, 1972; Achalabhuti and others, 1975). Malaysia is a significant world producer of ilmenite based on this byproduct recovery (Table 3). The second example is in the San Gabriel Mountains of California, where mining is primarily for sand and gravel. Ilmenite has been recovered recently from these deposits (Industrial Minerals, 1986). Recovery of titanium oxides has been evaluated for a number of other deposits of fluvial sand and gravel or silica sand (Davis and Sullivan, 1971; Gomes and others, 1979, 1980; Force, 1980a). Ilmenite was once recovered from fluvial placer deposits of monazite and other minerals in Idaho (Storch and Holt, 1963).

ECONOMIC PROGNOSIS

Sierra Rutile has demonstrated the potential profitability of fluvial rutile deposits. Far too little is known about the distribution of rutile-rich crystalline rocks and their erosional debris to claim that this deposit is unique. Where shoreline placer deposits of rutile become depleted, exploration in adjacent valleys may lead to fluvial rutile deposits.

For fluvial and glaciolacustrine ilmenite deposits, the lack of ilmenite alteration is a problem shared with magmatic ilmenite deposits. Nonmarine ilmenite placers might compete where unweathered ilmenite can be utilized, and if the economic recovery balance between high-grade hard rocks and low-grade disaggregated sediments shifts in favor of the sediments.

METHODS OF EXPLORATION

Exploration for nonmarine placer deposits of titanium minerals must center around the lithology of source rocks, as all such deposits known are proximal to and closely reflect their sources. A rutile-bearing source is far more promising than an ilmenite-bearing source, as nonmarine placers seem not to contain altered ilmenite. A sedimentary source containing already-weathered ilmenite may be favorable, however.

For fluvial deposits, deep weathering of the source terrane is an additional requirement implied by our present knowledge of such deposits. In some deposits, this weathering is as important as fluvial placer concentration in upgrading debris from the source.

Unusual geomorphic situations are required in fluvial systems to prevent the dilution of favorable by unfavorable debris (Fig. 59). Exploration should be concentrated where these situatins permit the accumulation of large-volume deposits derived from a favorable source terrane.

Not as much is known about favorable paleodispersal patterns in Pleistocene glaciolacustrine systems, because of the complication of subglacial transport. However, known deposits are near their source. The presence of thick deltas in partially filled basins of former glacial lakes may be obvious from topography or known from geologic maps showing Quaternary units.

The tendency of fluvial deposits to become richer at depth makes exploration with motorized equipment necessary. Favorable lithologies are coarse and medium sands with little gravel or mud matrix; axial-channel deposits may be more favorable than overlying point-bar deposits.

Geophysical exploration techniques for these deposits are not well known. Techniques that are successful in shoreline placer deposits, discussed in the next chapter, may be useful with nonmarine deposits.

Chapter 9.

Placer deposits in shoreline-related sands of Quaternary age

Titanium-mineral deposits in sand bodies on Quaternary shorelines are currently of greater importance than deposits of any other type. Even magmatic ilmenite deposits rank a poor second in importance behind shoreline placers. In the United States, virtually all current production and 36 percent of all identified resources are contributed by this deposit type (Table 5). The equivalent figures for the entire world are about 55 percent (1987 production) and 45 percent (Table 4), respectively. Individual shoreline placer deposits can contain tens of millions of tons of titanium oxide minerals.

Shoreline placer deposits acquired considerable economic stature in the early days of the titanium industry, during the late 1940s, and their importance increased with development of the chloride process. Only the shoreline deposits have been able to supply the high-TiO_2 altered ilmenite favored for that process; in addition, shoreline deposits commonly supply rutile, which is even more valuable. The advantage conferred by a loose, well-sorted sediment with uniquely attractive mineralogy is difficult to counter, even with a rock of much higher TiO_2 grade. The currently economic rutile "placer" deposits of eastern Australia, for example, have cutoff grades well below the average crustal abundance of TiO_2 (about 1.4 percent).

The complexity of shoreline placer deposits is far greater than meets the eye (Fig. 58B). Other chapters treat hidden factors such as source terranes (Chapters 2 and 3), the weathering history (6), fluvial conduits (8), and hidden components of the concentration system (7) necessary for the formation of these deposits. In addition, the shoreline sands embody a history of sea-level stands and sediment budgets and represent a number of individual depositional and preservational environments. Thus, the study of shoreline placer deposits represents a culmination of several disciplines and requires all the skill that any investigator can muster.

GENERAL CHARACTER

Shoreline deposits of titanium oxide minerals show a great deal of variation, as one might expect from the large number of factors involved. Yet the most valuable deposits, that is, those in which all the factors are optimal, show a great number of features in common.

The deposits (of Quaternary age) are found on trailing margins of continents, at latitudes lower than 35°, and are fed by detritus mostly from high-grade metamorphic source terranes. Host sands form coast-parallel surficial bodies, with basal elevations that generally correspond to local Quaternary sea-level high stands. Many of the sand bodies form topographic highs that represent former barrier islands and/or eolian dunes. Sand bodies containing economic titanium-mineral deposits are typically 10 m thick, 1 km wide, and more than 5 km long. Amalgamated smaller bodies of similar origin may be economic, as may superimposed composite bodies of different facies. Sets of sand bodies that are parallel but not contiguous commonly represent the strandlines on flights of marine terraces.

The sands of such bodies are medium- to fine-grained, well sorted, and generally well rounded. Typically the sands are unindurated. Depositional facies vary as discussed below. Well-developed weathering profiles are superimposed on most sand bodies. Local cementation by iron oxides, clay, and/or humate commonly marks postdepositional geochemical interfaces (such as former water tables) within deposits.

The heavy-mineral assemblage is restricted to species resistant to weathering and consists largely of the economic suite altered ilmenite, rutile, zircon, aluminosilicate minerals, and monazite. The proportions of these minerals vary considerably, and with these proportions varies the total heavy-mineral content

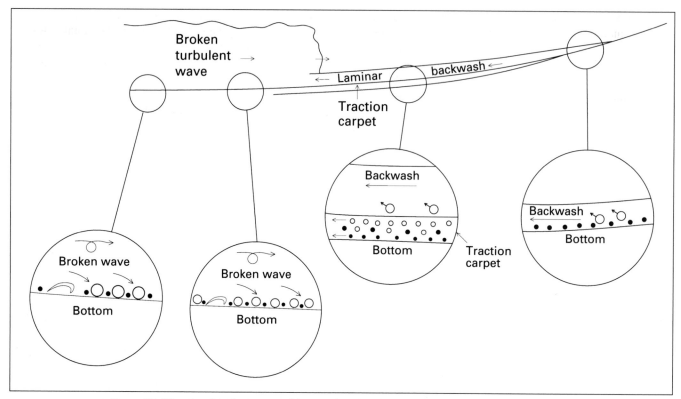

Figure 62. Diagram of various stages of heavy-mineral enrichment in the swash zone of a beach. A breaking wave is decelerating from left to right, whereas the accelerating backwash of the previous wave is moving from right to left. Insets show two stages of deposition from the turbulent breaking wave by suspension equivalence (toward the left), and two stages of placer enrichment by superimposed entrainment and transport equivalence (toward the right). Heavy minerals shown dark.

necessary to make a deposit economic. The range of total heavy minerals in today's economic deposits is from less than 1 percent to more than 25 percent; thus, the color of ore sand varies from white to black.

PROCESSES OF ENRICHMENT

General principles of placer enrichment, discussed in Chapter 7, are applied here to the particular environment of the beach face. Indeed, this is the environment in which such principles are most clearly illustrated. Except for attributed statements, the following observations are my own.

Detritus supplied to a beach face is commonly somewhat sorted by settling velocity (Komar and Wang, 1984). However, the main separation of particles on the basis of their settling velocities takes place on the upper part of the beach face, the swash zone. A breaking wave carries a charge of turbulent sediment-laden water from the lower, submerged beach face onto the swash zone (Fig. 62). The wave decelerates as it advances. Grains are deposited from turbulent suspension and spread across the surface of the swash zone, as functions of their settling velocities. Grains having the same settling velocities are deposited together. The grains that are deposited first, at the bottom of the swash zone, are those having the highest settling velocities. This

results in a continuous but narrow spectrum of grain sizes for each mineral, with finer grains deposited toward the top of the swash zone (Miller and Zeigler, 1958). At any given place, coarser light minerals are deposited with finer heavy minerals. At the upper limit of wave swash, water motion either ceases or is slow and along shore, and most of the sediment load is dropped.

When backwash begins and accelerates downslope, the water has little suspended load and therefore has an enhanced erosive capacity. Water motion in the backwash is a sheet flow, unlike the turbulent breaking wave. In the viscous boundary layer of this flow, velocity and thus erosive capacity are functions of distance above the bed. The larger grains, which project into the current, are plucked from the bed, and smaller grains are left behind. Because of prior differential settling from suspension, the removed coarser grains are light minerals. Thus, in the upper portion of the swash zone, heavy minerals are concentrated on the surface of the beach by the law of entrainment equivalence (of Slingerland and Smith, 1986).

Farther down on the swash face, an entire carpet of grains has been entrained in the backwash and is transported down the swash zone as traction load in the flow. Finer-grained dense minerals ride low in this carpet and travel more slowly than the coarser light minerals above, in transport and dispersive equivalence (of Slingerland and Smith, 1986). Thus the lower part of the swash face commonly contains a thin, shallowly buried heavy mineral concentration after the return of backwash to the surf zone (Clifton, 1969). This concentration can be traced in "subsurface" to the surficial concentration on the upper swash face. The heavy minerals of the lower subzone may be less dense than those of the upper swash face, however (Komar and Wang, 1984), because of slightly coarser equilibrium grain size.

In the lowermost swash face, the return flow may eventually become more turbulent, ending separation by grain size and heavy-mineral enrichment. This subzone is commonly marked by a line of shell lags and pebbles that begins the coarser deposits of the lower beach face (cf., Clifton and others, 1971). The buried heavy-mineral lamination terminates at coarse sediment of this subzone.

Particles left by the backwash of one wave are roughly in equilibrium with the next breaking wave, as the grain population left by each is the same except for the plucking of coarser constituents by backwash. The overburden of the buried heavy-mineral lamination is preferentially exposed to any erosion by succeeding waves.

Numerous authors have noted that rich beach concentrations contain disproportionate percentages of the densest minerals. This is due to sorting by density and grain size within the heavy-mineral population on beach segments where most light grains have already been removed.

Enrichment processes on the beach face operate most effectively during storms or other periods of high wave energy. During these storms, high onshore winds transport sand from the beach and deposit it in eolian landforms above high tide. Heavy minerals are transported preferentially, because the upper swash zone, where heavy minerals are exposed, is driest and least cohesive. Thus, storm periods are optimal not only for heavy-mineral concentration on the beach face but also for storage of the concentrate in the eolian environment. When fair weather returns, the dune deposits can be richer in heavy minerals than newly accreted fair-weather beach deposits, which bury the enriched beach deposits.

Thus, dune deposits are enriched in heavy minerals not only by concentration that occurs there but also by a highly selective supply system. Bulk enrichment of heavy minerals in dunes relative to adjacent beach deposits probably does not occur; studies by Shideler and Smith (1984) and Bradley (1957) that claim eolian enrichment were based solely on surficial samples of summer beaches. That is, they did not sample the higher-energy system that supplied the dunes. Authors who have studied whole beach-dune systems (Neiheisel,1958; Gillson, 1959; Lissiman and Oxenford, 1975; Welch and others, 1975; Collins and Hamilton , 1986) agree that the heavy-mineral contents of coastal eolian deposits are less than those of adjacent beach deposits and that eolian heavy-mineral assemblages contain smaller proportions of the densest minerals. Others have noted that lower dunes have greater heavy-mineral contents than higher dunes (Neiheisel, 1958; Fockema, 1986).

Heavy-mineral concentration does occur in the eolian environment, mostly on the scale of individual bed forms. Heavy-mineral–rich laminae form in response to daily variations in wind speed that change the shapes of eolian bedforms (Hunter and Richmond, 1988).

FORMATION AND PRESERVATION OF WHOLE DEPOSITS

Heavy-mineral concentration as discussed in the preceding section is a process involving lag enrichment on the swash face. The dominance of erosion over deposition on the swash face ensures that backwash efficiently sorts the available material, to produce a layer enriched in fine dense minerals. Thus the process of erosion is essential to heavy-mineral concentration. We shall see that many individual enriched layers lie on unconformities of minor to major significance.

Yet a typical economic deposit as described above contains millions of cubic meters of sand and is clearly the result of preservation of countless superimposed concentrated layers. The clue to the apparent paradox in the roles of erosion and preservation comes from the geometry of enriched layers. In this context, there are two types of shoreline heavy-mineral deposits: those in which heavy-mineral concentration occurred at the depositional site, and those in which enrichment occurred elsewhere. These can be referred to as in-place and transported enrichments.

In-place enrichments

Many in-place enrichments show progradation of swash face environments at constant sea level. Heavy-mineral–rich layers in these deposits show an imbricate arrangement of former shore faces that dip seaward in a progradational package. Each layer enriched in heavy minerals represents an erosional change in beach profile to a storm-influenced configuration, and the overlying low-grade layer represents subsequent burial by fair-weather deposits. Where progradation of fair-weather deposits is sufficiently great, they will protect the deposits of one storm from the next storm, and the younger storm profile will be seaward of the older (Fig. 63). Examples of strandline enrichments formed by progradation at nearly constant sea level are those of eastern Australia, formed predominantly by progradational burial of storm concentrations (Fig. 64A). Some of the heavy minerals in these deposits were supplied by erosional reworking of older shoreline deposits.

Deposits on the western coast of Australia show progradation amplified by local tectonic uplift. Each concentrated layer is related to a slightly lower base level than the preceding layer (Figs. 64B, 64C). This is a second type of preservation of in-place enrichments; falling sea level protects previously enriched deposits from marine erosion.

A third type of preservation of in-place enrichments on the swash face forms at extreme high tides of meteorological origin.

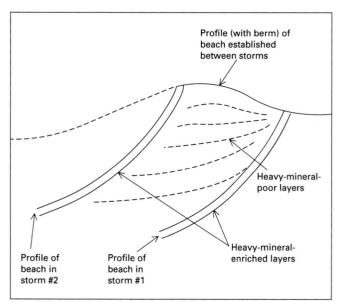

Figure 63. Diagram of preservation by progradation of heavy-mineral–enriched layers produced by storm erosion.

Lag enrichments formed on swash faces at these high levels are protected from subsequent marine erosion and may be buried by later eolian deposits.

All three presented examples of the preservation of in-place enrichments involve beach deposits. In-place enrichments probably also occur in other types of shoreline environments, such as tidal inlet deposits. These have been studied little.

Transported enrichments

Enrichments may be reworked from the environment where enrichment occurred and fed in bulk to a new environment where they can better be preserved. The most important of these new environments is eolian. In eolian deposits, an enrichment is removed grain by grain from the swash zone where enrichment occurred to a new environment with greater stability and storage capacity. Eolian deposits may acquire great volume. Transported enrichments may also occur in other environments, such as washover fans.

FACIES

Beach deposits

Many of the world's more important placer deposits of titanium minerals are true beach deposits, and every major district of shoreline titanium-mineral deposits includes some beach deposits. Many deposits (such as that shown in Fig. 65) are composites of beach sands and overlying eolian sands, in which the beach-facies deposits are typically higher in grade. Thus it would be difficult to calculate the relative magnitude of beach and eolian titanium-mineral resources.

The more valuable beach deposits include both Pleistocene and Holocene examples. Even the Pleistocene deposits commonly preserve strong elements of their original physiography and apparently represent high stands of Pleistocene sea levels. They are found on marine terraces roughly parallel to the present shore. Economic beach placer deposits represent several types of beaches, such as barrier islands, spits, and cliffed shorelines. Progradation plays an important part in preservation of mineable deposits regardless of beach type.

Beach-facies deposits in cross section consist essentially of alternations of sands poor in heavy minerals and sands that are enriched (Fig. 64). Enriched intervals may be as much as several meters thick, in which case, mining ventures can focus on individual concentrations. More commonly, individual concentrations are about a centimeter thick, and progradational sequences containing many thin concentrations are mined as if the deposit were disseminated.

Individual seaward-dipping heavy-mineral laminae typically represent swash-zone profiles under storm-wave conditions (Fig. 63). On the tectonically stable coast of eastern Australia, the bases of Holocene placer enrichments are found to be at modern mean sea level (Fig. 64A). Thus, enrichment appears limited to former swash zones. In depositional packages representing progradation of the entire beach face, which may be 10 m thick or more, the portion showing greatest concentration should be an upper interval representing the swash zone.

In map view, the locales of heavy mineral concentration may be highly localized within a shoreline complex having low heavy-mineral content (McKellar, 1975; Force and others, 1982). Most segments of most beaches are not efficient "concentrators." Particular portions of shoreline compartments may systematically show the greatest enrichment. This topic is further discussed under exploration methods.

Eolian deposits

Coastal eolian dunes contain significant portions of the titanium-mineral resources in shoreline deposits of the United States, South Africa, and Australia, that is, those districts where mechanized mining can be of low-grade material. Their great volume and homogeneity make some eolian sand bodies economic even at low grades. Composite eolian dunes, such as Fraser Island, Australia, and Trail Ridge, Florida-Georgia, can contain several billion tons of sand, though only a fraction is ore grade. Other eolian sands, such as those of Richards Bay, South Africa, are high-grade deposits.

The coastal dunes containing heavy-mineral resources are both Pleistocene and Holocene in age. The dunes can be subdivided into three main types: foredunes, transgressive dunes, and stationary dunes. Table 17 gives examples of deposits of each type.

Foredunes are those immediately adjacent to beach deposits. They form the backbone of many barrier islands and the ridges in many accretionary beach-ridge complexes. Foredune deposits

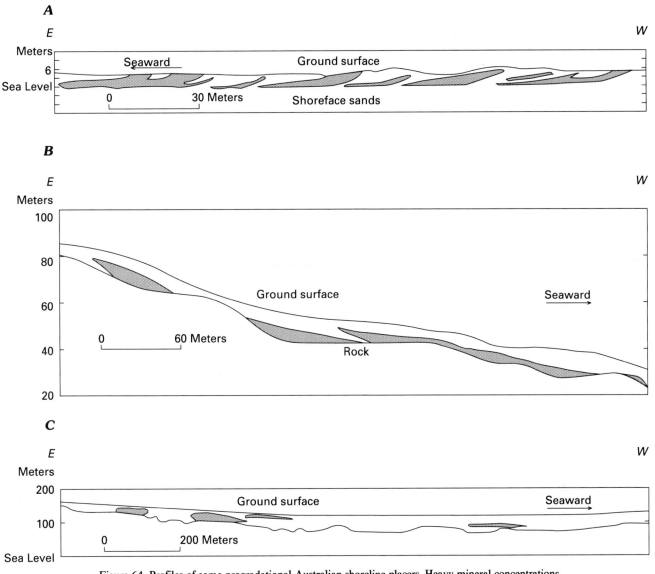

Figure 64. Profiles of some progradational Australian shoreline placers. Heavy mineral concentrations shaded. The scales differ but vertical exaggeration has been eliminated. A. Cudgen, after McKellar (1975). B. Yoganup, after Baxter (1982). C. Eneabba, after Lissiman and Oxenford (1973).

may be interbedded with true beach deposits formed during extreme high tides. Mining of some foredune deposits probably occurs in virtually all beach-facies deposits. In some Holocene deposits of eastern Australia, separate mining of beach and foredune facies was practiced when operations were smaller in scale.

Transgressive dunes are those that have become detached from beach deposits and have migrated inland. Deposits in transgressive dunes commonly overlie lagoonal or swamp deposits (Thom and others, 1981; Force and Garnar, 1985; Force and Rich, 1989). Transgressive dunes can be further subdivided into longwall and cliff-top dunes (Thom and others, 1981; Short, 1987, 1988). Longwall dunes remain parallel to the shore as they migrate inland over topography of low relief. Many economic heavy mineral deposits are of this type (Table 17). Cliff-top dunes

are characteristically parabolic, composed of sand transported up eolian ramps (commonly later eroded) on cliffed or other steep shorelines. The Jennings Eneabba heavy mineral deposit of Western Australia apparently formed as a cliff-top dune.

Stationary dunes are tied to bedrock features. They probably evolved from other dunes but in later stages of development accumulated successive additions of transgressive dunes without movement. Those of the southern Queensland coast in Australia are collages of successively accumulated large parabolic transgressive dunes, separated in cross section by buried soil horizons (Thompson and Ward, 1975; Ward, 1977, 1978).

Migration of transgressive dunes preserves slip-face laminae, which thus characterize most of the dune interior (Bigarella and others, 1969). Heavy-mineral-rich fine laminae outline the

Figure 65. Annotated photograph showing eolian dune (d) overlying regressional beach deposits (b), both of Pleistocene age, together forming ore being dredged in Tomago area near Newcastle, New South Wales. Nearby, 12,000-year-old freshwater peats intervene between the two facies of deposits (Thom and others, 1981).

former slip faces in some economic deposits (Macdonald, 1983; Force and Garnar, 1985; Force and Rich, 1989). In other eolian deposits, heavy-mineral distribution is more nearly disseminated.

SEDIMENT DISTRIBUTION

The supply and distribution of sediment within a coastal compartment can be an important control on the formation of titanium-mineral deposits. Commonly, sediment of more than one type is supplied, and only one has the mineralogy and grain size to form a valuable deposit. In economic deposits, sediment distribution is such that most of the promising material is fed to a "concentrator" such as the swash zone, whereas most of the rest of the material is deposited elsewhere. The mechanisms for accomplishing this are as varied as are source areas, conduits, and depositional environments and defy orderly description. Some examples are given in descriptions of individual districts.

Changes in the distribution and type of sediment with changing sea level are particularly common. For example, drowning of river mouths or individual headlands may activate different distribution systems. Such changes are recorded as mineralogic differences in strandlines at different elevations within the same coastal compartment. Cyclic sea-level changes may cause deposits formed at different times but at the same sea level to have comparable mineral assemblages.

Headland erosion of friable sandstones is apparently an important factor in changes of sediment distribution in several districts described in this chapter. At high sea levels, the drowning of river mouths cuts off fluvial supply of immature mineral suites, and headland erosion becomes a larger component of mineral supply. Where friable sandstones containing valuable mineral

assemblages (at low concentrations) constitute the headlands, high sea level activates the concentration of this valuable mineral assemblage on beaches.

TIME SERIES OF DEPOSITION, AS RECORDED ON FLIGHTS OF MARINE TERRACES

In some coastal compartments, a series of local high stands of Quaternary sea level is represented by contemporaneous shoreline deposits and related facies. These deposits are preserved on marine terraces at elevations ranging from sea level to tens of meters and inland as much as tens of kilometers from the present shore. Flights of such terraces are well illustrated in the southeastern United States, where as many as six shoreline deposits can be recognized in South Carolina (Colquhoun, 1965; Thom and others, 1972; McCartan and others, 1990) and Virginia (Oaks and others, 1974; Force and Geraci, 1975). In some places, former barrier segments, capes, and tidal inlets can be recognized from the shapes of shoreline sand facies, and these features may closely mimic present features of the same coastal compartment (as in Fig. 66).

Each shoreline deposit of a set may contain heavy-mineral enrichments (Baxter, 1977; Force and others, 1982). These deposits typically differ in the extent of postdepositional weathering (Chapter 6). Weathering differences make the older, more elevated deposits of each coastal compartment potentially the more valuable. Mineral assemblages may also vary, because of sea-level controls of distribution patterns.

DEPOSITS NOW OFFSHORE

During most of the Quaternary, sea level has been lower than it is now (Fig. 55). Any sea-level stand, especially a still-

TABLE 17. TITANIUM-MINERAL DEPOSITS IN DIFFERENT TYPES OF QUATERNARY SHORELINE SAND BODIES

Type	Example	
	Holocene	Pleistocene(?)
Beach deposits	Cudgen, East Australia Minninup, West Australia Travancore coast, India	Green Cove Springs, Florida; Jerusalem Creek, East Australia; Yoganup, Capel, and Eneabba, West Australia
Eolian deposits		
Foredune	Cudgen, East Australia	?
Transgressive	Williamtown, East Australia Richards Bay, South Africa	Trail Ridge, Florida; Bridge Hill, East Australia; Jennings Eneabba, West Australia
Stationary		Stradbroke, Moreton, and Fraser Islands, East Australia

stand, can leave behind a shoreline sand deposit, and most of these should now be present on the continental shelf. By this reasoning, the continental shelf should be favorable for shoreline placer deposits. Several factors complicate this hypothesis (Attanasi and others, 1987). For example, transgression commonly destroys the upper portions of shoreline sand bodies (Swift, 1968); these are the portions containing heavy-mineral enrichments. Constituent heavy-mineral grains may be transported landward (possibly as transported enrichments) or may be incorporated in reconstituted sand bodies on the continental shelf. All shoreline sands formed at sea levels lower than today's have subsequently suffered transgression.

Another limitation of offshore sands is their weathering histories; shoreline sand bodies now offshore formed at times of slow weathering and have been submerged through most of their history. Thus they should be less weathered than their onshore counterparts (Chapter 6).

Despite theoretical limitations, sand bodies on continental shelves locally contain substantial quantities of valuable heavy minerals (Grosz and others, 1986; Grosz, 1987). On the continental shelf of the eastern United States, surficial heavy-mineral enrichments are numerous. Vibracore drilling shows that appreciable volumes of heavy-mineral–bearing sands are present in some areas.

Off a few coasts, shoreline sand bodies have survived transgression and are partly buried on continental shelves, with original morphology partially preserved (Flemming, 1981; Schluter, 1982). In these bodies, titanium oxide minerals should be present approximately as in bodies above sea level. Off other coasts, sand bodies exposed on continental shelves are Holocene in age and formed in place (Duane and others, 1972). These bodies may be derived from older shoreline sands that were redistributed by transgression.

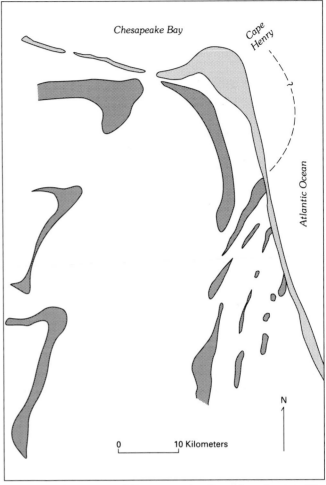

Figure 66. Map pattern of Quaternary shorelines of southeastern Virginia, after Force and Geraci (1975). Pleistocene shorelines shaded darker than Holocene ones.

CEMENTATION AND WEATHERING

Weathering of titanium minerals in shoreline deposits is extensively discussed in Chapter 6. Much of that discussion is aimed at discrimination of weathering that occurred before and after deposition of heavy-mineral grains. Cementation of shoreline sands is in part a result of postdepositional weathering and is of great economic importance.

The degree of cementation of permeable surficial sands of shoreline sand bodies varies with the age of a deposit and depth within it. Holocene deposits are little cemented (except locally by calcareous material); younger Pleistocene deposits, such as that shown in Figure 65, contain cements but are still friable. Older Pleistocene deposits, such as those of Eneabba and Trail Ridge, contain cements that can make a hammer ring.

The noncalcareous cements are apparently of two intergrading types. At one extreme are pure organic cements, consisting of amorphous black humin or humate. At the other extreme are pure iron hydroxides. Variable amounts of clay matrix can be incorporated in either type. Most cements are intermediate and contain large amounts of iron complexed with organic matter. Aluminum is similarly complexed in some deposits (Ward and others, 1979).

The cements generally form thin, tabular layers that represent the positions of present and former water tables (Cannon, 1950; Welch and others, 1975; Ward and others, 1979; Baxter, 1982; Thompson and Bowman, 1984; Force and Garnar, 1985). Massive cements can apparently form by amalgamation of these thin layers (Force and Rich, 1989). Humate cement forms at the water table where humic acids, derived from vegetation at the ground surface, are neutralized by groundwater. The iron component of the cements is derived from acid leaching of sediments above the water table, and iron also precipitates at the water table by neutralization (and perhaps oxidation).

The iron component of cements is related to ilmenite alteration. In many deposits, ilmenite above the water table has been intensely leached (Chapter 6). Iron in cements precipitated at the water table probably was leached in large part from ilmenite (Welch, 1964; Puffer and Cousminer, 1982). This hypothesis is supported by the preservation of less altered phases such as magnetite within iron-cemented layers (New Jersey: Jim Stern, personal communication, 1974; Western Australia: Lissiman and Oxenford, 1975; Baxter, 1986). Early-formed iron cements prevent further alteration of encased material.

MAJOR DEPOSITS

Jacksonville district, Florida and Georgia

The only titanium-mineral deposits currently mined in the United States are in northeastern Florida, near Jacksonville (Fig. 67A). The two active deposits, known as Trail Ridge and Green Cove Springs, are part of the Jacksonville district, which includes many deposits in northeastern Florida and southeastern Georgia. Mining was formerly conducted in the district near Folkston, Georgia, and Boulogne and Jacksonville, Florida. Unmined deposits of the district include the Cumberland Island and Altamaha Plantations deposits of Georgia and the Amelia Island and Yulee deposits of Florida. All these deposits are Pleistocene, related to former shorelines on at least three different marine terraces (Fig. 67B). Modern beach deposits of the district were also mined from 1916 to 1929. Altogether, the district has produced about 5×10^6 metric tons of TiO_2 and still contains resources of about 14×10^6 metric tons of TiO_2 (Force and Lynd, 1984).

The district is the southern end of the Atlantic coastal plain, which to the north contains additional subeconomic deposits (Force and Lynd, 1984). The district has little vertical relief and is mostly wooded. Surficial units are all late Cenozoic in age, but exposure is extremely poor. Mining in the district is entirely by dredge, so that geologic relations in the mines are obscured. As a result, geologic information on the deposits is limited to drill cuttings and a few cores, with the exception of Trail Ridge.

Trail Ridge deposit. Sands of Trail Ridge are mined by DuPont at two dredging operations. Altered ilmenite and zircon are the main products and dominate the heavy-mineral assemblage. Heavy minerals average about 4 percent of fine- to medium-grained sands that form a surficial sand body up to 20 m thick. This sand body forms a ridge about 2 km wide and more than 160 km long, extending well into Georgia. Only the southern 29 km contains known ore-grade sands (Fig. 67A). The elevation of the ridge crest is about 75 m toward the southern end and about 45 m toward the north; the base of the sand shows a similar slope (Fig. 67C).

The Trail Ridge deposit has been studied by Spencer (1948), Pirkle and others (1970, 1971, 1977), Pirkle and Yoho (1970), Force and Garner (1985), and Force and Rich (1989). Force and Rich include an extensive review of other previous work.

Temporary exposures (caused by lowering of the dredge-pond surface) of ore in vertical profile have shown Trail Ridge to be a great eolian dune (Force and Garnar, 1985). The form and

geologic relations of this dune indicate that it is of the transgressive type. Preserved bedding is that of slip faces dipping steeply southwest.

The ore sand at Trail Ridge has a convex exposed upper surface but a subhorizontal base and overlies a 1.5-m lignitic peat unit in the area of current mining. This peat is of freshwater swamp origin (Rich, 1985). Probably the swamp was impounded by the dune. Sand impurities of the peat record the approach of the transgressive dune, which subsequently overrode the swamp (Force and Rich, 1989).

The exposed upper part of the ore sand consists of several subhorizontal humate-cemented layers, overlain by a bleached zone (Fig. 47). The humate layers intersect slip-face bedding at a high angle and are present in a zone up to 3 m thick. The bleached zone, 2 to 3 m thick, is rootlet mottled and contains heavily altered ilmenite. The humate layers apparently represent precipitation by humic acid neutralization at the water table and former water tables.

The bulk ore sand has a median grain size of about 0.3 mm, is well sorted (So 1.16 to 1.32 for the sand fraction alone, So 1.23 to 1.35 if clay matrix is included), and is positively skewed. Constituent grains are well rounded, and even prismatic minerals show appreciable sphericity (Pirkle, 1975). Grain surfaces are frosted, analogous to eolian sands elsewhere (Force and Rich, 1989). Heavy minerals are slightly finer than light minerals (Fig. 60C). The heavy-mineral assemblage includes altered ilmenite and leucoxene (about 50 percent [%]), zircon (15%), staurolite (15%), sillimanite (5 to 6%), tourmaline (5%), rutile (2 to 3%), and kyanite (3%), with minor spinel, corundum, and monazite.

In detail, two distinct sand lithologies are present (Force and Garnar, 1985; Force and Rich, 1989; Fig. 49). Slip faces are outlined by alternations of predominant light-colored sand and subordinate dark-colored sand. The dark-colored sand is far richer in heavy minerals and is finer grained in both the light-mineral and heavy-mineral fractions. The dark laminae preferentially contain the denser heavy minerals of the deposit, as fine grains, whereas staurolite, tourmaline, and sillimanite are partitioned toward the light-colored layers as coarser grains (Fig. 68).

No coeval beach or marine facies have yet been found associated with Trail Ridge, although marine facies overlie Trail Ridge sands toward the north (Pirkle and Czel, 1983). Analogy with Australian transgressive dunes suggests that such coeval marine facies were probably present within several kilometers to the east. Figure 69 depicts the probable relation of shoreline, wind direction, and dune orientations.

The age of the Trail Ridge deposit is still unknown. Microflora of the peat show it to be post-Miocene, while ^{14}C dates show that it is older than 45,000 years. Thus, Trail Ridge sands could be either Pliocene or Pleistocene.

Trail Ridge sands have been a focus of much of the work on weathering of ilmenite. Pirkle and Yoho (1970) and Temple (1966) documented a great increase in leucoxene at the expense of ilmenite in the surficial leached zone, above the water table (Fig. 47). This weathering is clearly postdepositional. Grey and

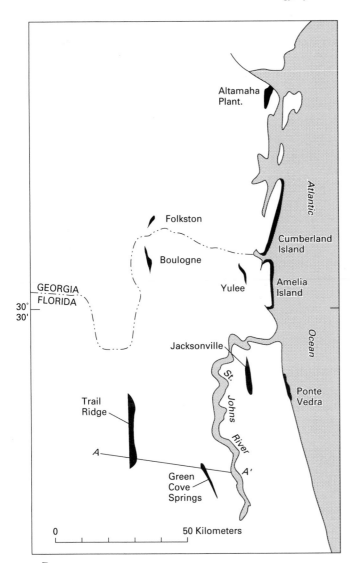

A

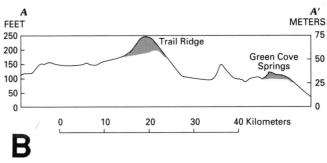

Figure 67. Titanium-mineral deposits of the Jacksonville district, Florida and Georgia. A. Map showing distribution of deposits (black) and line of cross section. B. Cross section of the Trail Ridge and Green Cove Springs deposits (shaded). C. Projection of deposit altitudes onto a north-south vertical plane, showing location of marine fossils overlying the flanks of the Trail Ridge dune. Data from Pirkle and others (1974) for Green Cove Springs and from Pirkle and Yoho (1970) and Pirkle and others (1971, 1977) for Trail Ridge.

B

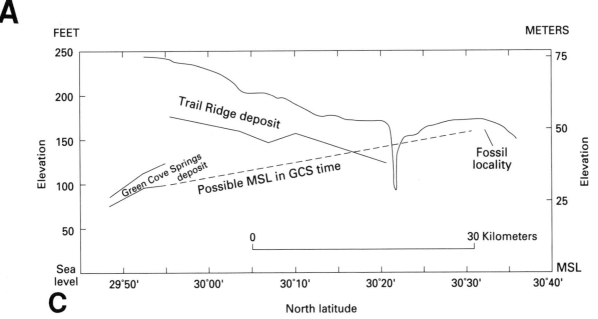

C

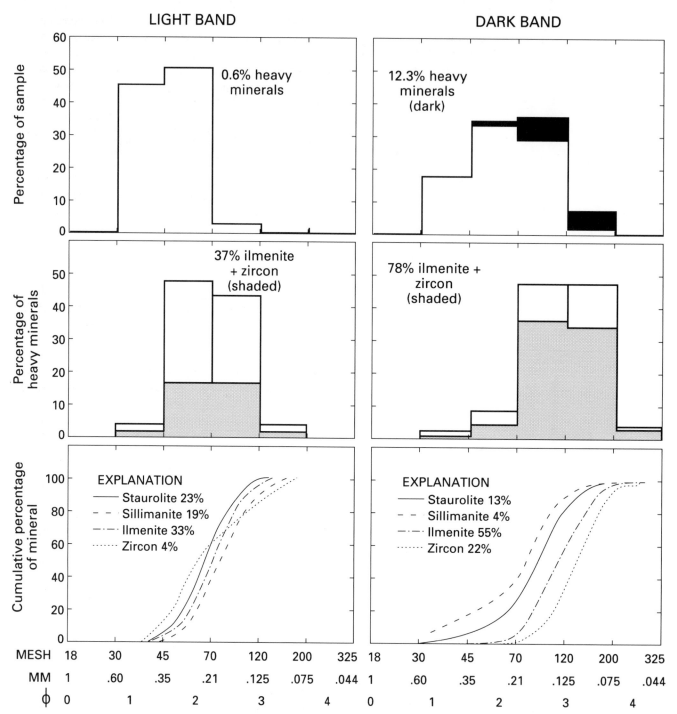

Figure 68. Size-frequency distributions comparing adjacent laminae rich and poor in heavy minerals at Trail Ridge. The upper graphs show histograms of the whole sample, the middle graphs show histograms of the heavy-mineral fraction, and the bottom graph shows cumulative curves for individual heavy-mineral species.

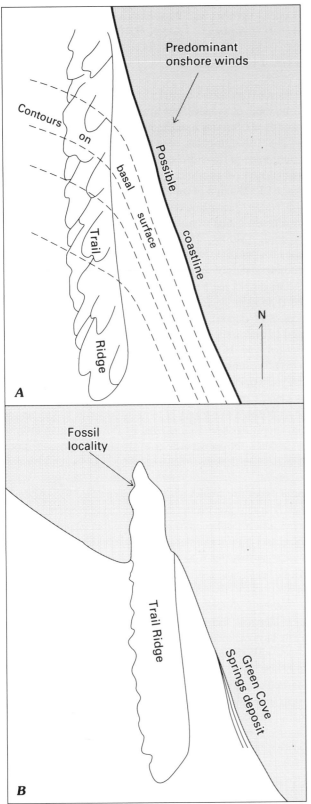

Figure 69. Schematic map views of former eolian system of Trail Ridge. A. Time of dune formation. B. Later time, with sea level taken from Fig. 67C.

Reid (1975) noted that ilmenite beneath the water table is also altered and proposed that a first stage of postdepositional ilmenite alteration occurred there. Sand morphology suggests another solution: grains below the water table were partly weathered at the time of deposition. Detrital oxides there show mixed alteration states but uniform high polish, acquired during deposition. Thus, this zone is a detrital mixture of variably altered minerals. Predepositional weathering is also suggested by the mineralogy of the eolian sand in the sandy peat facies underlying the ore, which contains variably altered ilmenite much like that of ore (Force and Rich, 1989). In the reducing peat environment, below the water table, first-stage alteration by oxidation was effectively prevented, so the mixed alteration assemblage represents the original detrital assemblage. Weathering of Trail Ridge sands thus took place in at least two stages, one before deposition and the other afterward.

The intermediate sources of Trail Ridge sands have been investigated by Pirkle (1975), who concluded that local upper Cenozoic formations could have supplied the material. The coarse grain size, mineralogy, shape, and surface features of Trail Ridge sand grains greatly limit the possible intermediate sedimentary sources, as well as immediate beach-facies sources.

Green Cove Springs Deposits. East of Trail Ridge and at slightly lower elevations (Fig. 67A) are the deposits at Green Cove Springs, now operated by Associated Minerals. These deposits formed in a progradational beach system. The most complete published descriptions, by Pirkle and others (1971, 1974), are rather limited in scope and do not include descriptions of subsequently discovered satellitic orebodies.

The main orebody is a progradational beach ridge complex oriented SSE-NNW, and is about 15 km long and 1 km wide. At the northern end, the elevation of the base of the ore is 98 ft (30 m); apparent southward tilting is discussed below. The ore is a surficial unit, normally about 6 m thick; it is a gray sand with humate-cemented layers, overlying brown sand in which heavy-mineral content decreases with depth. The thickness of the gray and brown sands as a whole is 13 to 17 m. If the sand is a single depositional sequence, it probably represents shoreface deposits overlain by swash-zone deposits.

The ore sands are fine, with mean grain size of about 0.17 mm. Trask sorting (So) values calculated from Pirkle and others (1971, 1974) are 1.3 to 1.4, almost as well sorted as Trail Ridge sands. Rounding of all minerals is moderate, but markedly less so than at Trail Ridge.

Heavy minerals constitute more than 3 percent of the sand and include altered ilmenite and leucoxene (about 50 percent [%]), zircon (about 15%), rutile (about 10%), staurolite and other aluminosilicates (about 15%), tourmaline, monazite (each about 0.3%), and, locally, epidote and garnet. Ilmenite concentrates contain 64 percent TiO_2. Mean grain sizes of these minerals are 0.10 mm for ilmenite and zircon, 0.13 mm for staurolite, and 0.08 mm for monazite (from data in Garnar, 1980). Rutile currently earns the most revenue (Fantel and others, 1986).

Other deposits of the district. Titanium-mineral deposits on shorelines roughly correlative with those of Green Cove Springs, at about 30 m elevation, formed at Boulogne (Pirkle and others, 1971, 1974) and Folkston (Fig. 67A). Shoreline sands at about 6 to 10 m elevation contain titanium-mineral deposits at Yulee (Pirkle and others, 1984), Altamaha Planatations, and Jacksonville. Younger Pleistocene deposits formed on Amelia and Cumberland Islands when sea level was only 2 m or less above present sea level.

Source and evolution. The ultimate sources of the heavy minerals concentrated in the deposits of the Jacksonville district are predominantly the high-grade metamorphic rocks of the southern Blue Ridge and inner Piedmont (Force, 1976b). Sedimentary hosts were intermediate sources for Trail Ridge (Pirkle, 1975) and probably for the other deposits of the district. Since the Piedmont-draining streams in the region contain a mineral suite far too immature (Neiheisel, 1976; Force and others, 1982) to have supplied the placers of the district without extensive predepositional weathering beneficiation, it is probable that this beneficiation occurred in the intermediate sedimentary hosts.

The immediate source of eolian sand supplied to Trail Ridge is still a problem. This source was both heavy-mineral rich and coarse grained. It seems likely that the dune was supplied by a beach to the east, but the Green Cove Springs beach deposit, the location of which seems appropriate, is heavy-mineral rich but too fine grained.

A suggestion by Pirkle and others (1974) that the base of the main Green Cove Springs orebody is tilted to the south is provocative, because the base of a progradational beach-ridge sequence records a horizontal surface at the time of deposition. A total of six boreholes show uniform southward tilt of about 0.5 m/km in the northern three-quarters of the main orebody (Fig. 67C). Michael Shepherd of Associated Minerals (written communication, 1987) warns that the base of the southern end of the orebody drops southward in discrete steps.

If one assumes that the Green Cove Springs body can be used as a tiltmeter, and that the Trail Ridge body is older but on the same crustal block, then the Trail Ridge deposit underwent the same southward tilting. At present the base of the Trail Ridge sand plunges northward at about 0.3 m/km, but before tilting, this land surface would have sloped about 0.8 m/km to the north (Fig. 67C). Eolian sand bodies like Trail Ridge commonly are deposited by climbing up sloping surfaces. Opdike and others (1984) based their uplift history of northern Florida on an assumption that Trail Ridge formed as a shoreline sand with a horizontal base. Instead it appears that the Green Cove Springs body is of shoreline origin and formed horizontal, but that Trail Ridge is eolian and never was horizontal. Both have subsequently been tilted down to the south. This may explain marine fossils overlying Trail Ridge in Georgia (Pirkle and Czel, 1983); the northern end of the dune ridge at the time the Green Cove Springs beaches were deposited could have had a drowned base (Fig. 67C, 69). Trail Ridge at this time could thus have been analogous to present high-dune islands off the southern Queensland coast of Australia.

Eastern Australia district

Rutile is present in coastal placer deposits of eastern Australia from south of Sydney (New South Wales) to north of Brisbane (Queensland), forming a single district more than 1,200 km long (Fig. 70). This district has been the world's most valuable single titanium-mineral resource, because of its great aggregate size and a unique mineral assemblage dominated by rutile, zircon, and ilmenite.[1] Production of rutile began in the 1940s near the New South Wales–Queensland border where first gold and then zircon were already mined from heavy-mineral concentrations of very high grade (Morley, 1981). Rutile production gradually expanded southward to the Hunter River area and northward to the high-dune islands of southern Queensland.

The district has been well described in terms of constituent mineral deposits, geologic relations among host sand bodies, and processes of deposit formation. Individual deposits have been described and mapped by Gardner (1955), Connah (1961), and Winward and Nicholson (1974). A summary by McKellar (1975) is the most recent for the whole district. Geology of Quaternary coastal sand bodies has been discussed and mapped by Roy (1982) and Thom (1983). Development of the bodies, including offshore extensions, in terms of sea-level history and sediment supply has been summarized by Roy and Thom (1981).

In detail, the district is diverse, as it includes both Pleistocene and Holocene deposits, each including both beach and eolian components. Commonly, more than one of the four resulting deposit types are mined in a single face (Fig. 65). Additional diversity is imposed on the district by being plastered against a geologically complex continental margin that includes three large sedimentary basins and three Paleozoic foldbelts (Figs. 70, 71). Considering that this is an embayed shoreline with hundreds of bedrock headlands, it is remarkable that a single general description can apply to the entire district.

General description and petrology. The Quaternary coastal deposits form discontinuous narrow coastal plains in discrete bays outlined by headlands, between the Tasman Sea and bedrock hills rising behind the beaches. Sands exploited in the district are most commonly medium- and fine-grained sands, quite well sorted and rounded. Median grain sizes are about 0.11 to 0.13 mm (Beasley, 1950). Quartz is the predominant light mineral; carbonate grains are commonly absent, and feldspar is exceedingly minor. Most of the shoreline sands of the district have very low heavy-mineral contents, less than similar sands from the east coast of the United States. Whitworth (1959) and McKellar (1975) estimate the ambient heavy-mineral content to

---

[1]Ilmenite had not been recovered in this district until recently because of the difficulty in separating it from minor chromite.

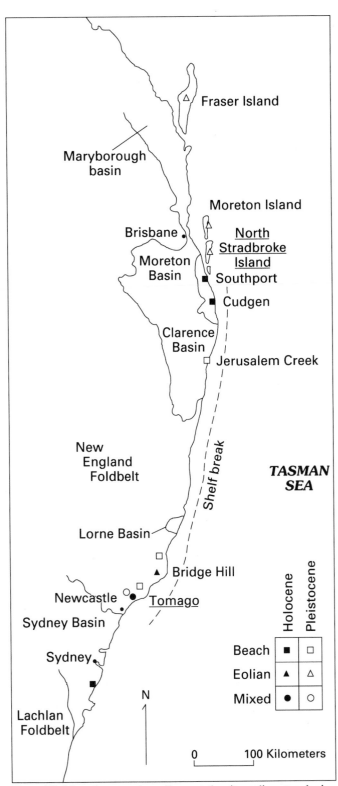

Figure 70. Map of eastern Australia coast showing sedimentary basins, intruded foldbelts, narrow shelf, and type and location of deposits. Underlined deposits in production.

be well under 0.1 percent. However, the mineral assemblage is so valuable that total heavy-mineral grades of less than 1 percent currently constitute ore.

In high-grade beds, rutile, zircon, and ilmenite make up greater than 90 percent of the heavy-mineral fraction throughout much of the district. Minor but characteristic members of the suite are tourmaline, monazite, chromite, and garnet. Cassiterite, epidote, magnetite, spinel, pyriboles, and metamorphic aluminosilicates are locally present. Colwell (1982a) noted that where heavy minerals are present at low concentration, the lighter heavy minerals form a larger portion of the suite. Beasley (1950) noted local increases of coarse noneconomic heavy minerals such as garnet around specific headlands, showing a contribution of headland erosion.

Figure 71 shows trends in rutile contents of heavy-mineral concentrates as a function of latitude through the entire district. Rutile distribution is fairly homogeneous but shows appreciable influence of differing adjacent geologic provinces; rutile contents are lower adjacent to the two southern foldbelts. Deposit age apparently is a less important control on rutile content; Hails (1969) reported little difference in mineral assemblage between Pleistocene and Holocene deposits. The limits of the district coincide not with lesser heavy-mineral concentrations but with changed assemblages; magnetite, ilmenite, pyriboles, epidote, and tourmaline suddenly become predominant to the south (Hails, 1969), whereas ilmenite gradually becomes predominant to the north (Connah, 1961). Hails (1969) reported abnormally high concentrations of andalusite, pyriboles, staurolite, and epidote in heavy-mineral assemblages in the central region of the district, adjacent to the New England foldbelt. My own coastal collections (included in Fig. 71) support his observation.

Electron microscope images of grain surfaces (Fig. 72) show that a given deposit is typically mixed in character. Quartz grains vary from angular to well rounded and polished. The *economic* heavy-mineral assemblage matches the more mature of these quartz populations in morphology.

Deposit types. Four types of deposits are exploited in the district; these are Holocene beach deposits, Holocene eolian deposits, Pleistocene beach deposits, and Pleistocene eolian deposits. Holocene beach deposits were recognized first and were the focus of most early mining; in fact, mining of the rich deposits of the immediately previous storm was an important component of this activity. Enrichment on beach faces is intensified by erosion because of seasonal changes in beach profile and reorientation of beaches (Beasley, 1948; McKellar, 1975). Thus, individual enriched beds lie on unconformities. Preservation of the enrichments occurs if they are protected from further erosion on the beach face (1) in those portions above normal high tide, (2) by rapid burial (progradation), or (3) by removal into the eolian environment. Individual layers of concentrated heavy minerals are (or were) as thick as 2 m, tapering and dipping seaward. In aggregate, these seams are en echelon, parts of progradational sequences with base level at about mean sea level (Fig. 64A). Beasley (1948) showed that little heavy-mineral enrichment in

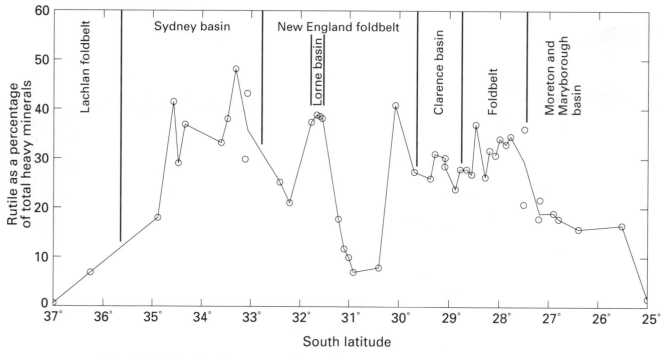

Figure 71. Variation in rutile component of heavy-mineral concentrates as a function of latitude through the eastern Australia district. Boundaries of geologic provinces at the coast are also shown. Values from Beasley (1950), Gardner (1955), Hails (1969, Fig. 6), McKellar (1975), and my own work. Duplicate samples at same location are averaged for curve. Data of Colwell (1982a) are not comparable and were not used.

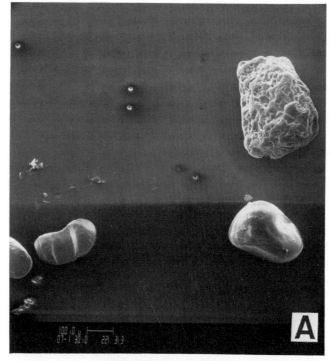
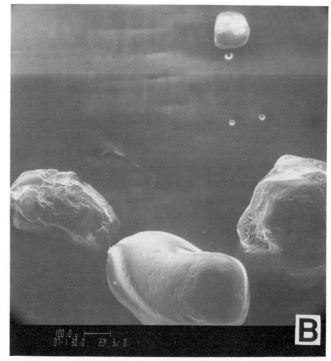

Figure 72. Scanning electron microscope photographs of grain morphologies of quartz and rutile from deposits in eastern Australia. A. Holocene beach sand deposit, Tuncurry. Two rutile grains on left, two quartz on right. B. Pleistocene eolian deposit, Tomago. One rutile grain above, three quartz below. Photographs by John Evans, U.S. Geological Survey.

Figure 73. Holocene eolian dunes in eastern Australia. A. Foredunes in the Jerusalem Creek area, looking north. Note the iron- and humate-cemented beachrock of Pleistocene age exposed in the beachface. B. Transgressive "longwall" dunes in the Williamtown area, looking east. The foreground is occupied by 4,300-year-old Holocene estuarine deposits (Thom and others, 1981).

these sequences extends below sea level. Supply of heavy minerals to the beaches is thought to be cannibalistic in part, as mining has depleted the total heavy-mineral stock of some coastal compartments, except where older heavy-mineral deposits are exposed in the beach face (Michael Shepherd, oral communication, 1986). Storm activity thus produces little placer concentration at present.

Pleistocene beach deposits differ little from their Holocene counterparts. They are also preserved as regressional beach plains containing seaward-dipping enrichments but are topographically more subdued (Thom and others, 1981). The sea levels represented by Pleistocene deposits are 0 to 5 m higher than those represented by Holocene deposits (Thom and others, 1981; Melville, 1984, Fig. 3). The mineral assemblages of Pleistocene and Holocene deposits are similar, probably because the mineral assemblages were already mature before deposition in most parts of the district and were little affected by further weathering. The main lithologic difference between Pleistocene and Holocene deposits is the cementation of the former by iron-aluminum hydroxides and humate (Fig. 73). The age of known Pleistocene deposits is about 140,000 years (last interglacial; Roy, 1982). Exploitation of Pleistocene beach deposits at Jerusalem Creek (McAuleys Lead) was among the earliest in the district; this deposit is along an unconformity between two sets of Pleistocene regressional beach plains. Pleistocene beach deposits are currently receiving a large share of the mining and exploration attention in the district.

Holocene eolian dunes, both as foredunes contiguous with beach deposits and as detached transgressive dunes moving inland (Fig. 73), commonly rest on Holocene beach deposits and are mined with them. A sizable resource in Holocene dunes is present in the Bridge Hill dune of the Myall Lakes area. The transgressive dunes may form shore-parallel ("longwall") features overlying estuarine deposits. In contrast to beach concentrations,

the eolian deposits contain disseminated heavy minerals at low grades.

Pleistocene eolian dunes represent a large share of the remaining resources in the district and are well represented in the Newcastle area (Fig. 65) and as the great high-dune islands (Stradbroke, Moreton, and Fraser) of southern Queensland. Like Pleistocene beach deposits, these dunes are cemented by humate and iron-aluminum hydroxides, down to the levels of present and former water tables (Thompson and Bowman, 1984). The Queensland islands are collages of dunes of various ages. Several stages of Pleistocene accretion of parabolic transgressive dunes are represented (Thompson and Ward, 1975; Stephens, 1982a). Individual dunes may have complex histories, shown by buried soil horizons. As the eolian sand extends well below sea level offshore from these islands (Kudrass, 1982), and as the dune formation in part occurred at times of low sea level, the islands are thought to represent transgressive dune complexes that were later drowned (McKellar, 1975; Ward, 1977, 1978).

Differentiation among these four types of deposits is locally possible based on petrography. In addition to the presence of iron oxide and humate cements, Pleistocene deposits have ilmenite fractions that are more uniformly altered than those of Holocene deposits. Ilmenite fractions of Holocene deposits are mixtures of slightly altered material and subordinate fresh ilmenite locally intergrown with other iron-titanium oxides. The eolian deposits contain some grains having eolian surface features such as frosting, but the variations among these grains are great.

Depositional setting. The embayed eastern coastline of Australia is wave dominated and faces a continental shelf less than 60 km wide. Predominant wave direction sets up a strong longshore drift to the north, but some headlands project offshore into the south-flowing Eastern Australia Current. Sand may be stored offshore south of such headlands (Kudrass, 1982; Field and Roy, 1985), or it may bypass a headland as eolian transgressive dunes. Present sea level has drowned the river mouths, and little immature fluvial sediment currently reaches the continental shelf (Roy, 1977; Roy and Crawford, 1977).

Quaternary geologic setting. The Holocene sands mined along the coast of eastern Australia are the shoreward portion of a sand sheet occupying the inner shelf. Throughout this sheet are found similar mineral assemblages, which are less sensitive to positions of (drowned) river mouths (Kudrass, 1982) than to adjacent geologic provinces (Fig. 71). The sand bodies contain a less mature assemblage in proportion to present depth either of water or burial (Jones and Davies, 1979; Colwell, 1982b; Kudrass, 1982; Reich and others, 1982; Stephens, 1982b). Thus, sands deposited at lower sea levels were less mature. Holocene sea levels have remained virtually constant for 6,500 years along this coast, permitting extensive reworking and progradation of shoreline deposits (Thom and Roy, 1985).

Pleistocene sediments on the continental shelf underlie the Holocene sands and form near-surface units on the midshelf (Jones and Davies, 1979; Schluter, 1982). These include several shoreline sand bodies. Petrography of the Pleistocene sands is a sensitive function of sea level, even more than for Holocene sands; the submerged Pleistocene bodies have immature heavy-mineral assemblages (Reich and others, 1982), comparable to those of related fluvial sands, whereas Pleistocene shoreline sands exposed above sea level have mature mineral assemblages comparable to those of exposed Holocene sands (Hails, 1969).

Interglacial periods (including the Holocene) and their high sea levels are represented by shoreline sands having mature mineral assemblages in present coastal regions; glacial periods and their lower sea levels are represented there by immature fluvial deposits. Near some present estuaries, immature Pleistocene fluvial deposits can be found stratigraphically between mature Pleistocene and Holocene shoreline sands (Roy, 1982).

Source rocks. Sources of minerals can be categorized as immediate, intermediate, and ultimate. The immediate derivation of beach and eolian deposits of eastern Australia is clearly from the cannibalism of older beach deposits, from offshore sands, and from sands being carried northward by longshore drift.

Authors also seem agreed on the intermediate sources—the Mesozoic sandstones of the Sydney, Clarence, and Moreton basins (Fig. 71), which have heavy-mineral assemblages dominated by rutile, zircon, ilmenite, and tourmaline, with minor chromite and tourmaline (Beasley, 1950; Gardner, 1955; McElroy, 1962; Galloway, 1972; Winward and Nicholson, 1974; McKellar, 1975; Davidson, 1982). Thus, the unique mineral assemblage of the shoreline deposits finds a match in an extremely unusual assemblage in Mesozoic sandstones.[2] The Hawkesbury Sandstone of the Sydney basin was probably the most important single source. The abrupt southern boundary of the district coincides with the southern end of the Sydney basin (Fig. 71). The New

England foldbelt coincides with low rutile values, punctuated by high values only near headlands of Mesozoic sandstone. This mineralogic tie to an intermediate source is a powerful tool in unraveling the developmental history of the deposits.

The ultimate source of the heavy minerals has long been debated. The most conveniently located possible sources are the New England and related Paleozoic foldbelts and their associated granites. Beasley (1950) recorded two rutile occurrences in the New England foldbelt and found rutile averaging less than 1 percent of heavy-mineral assemblages in streams draining the foldbelt. He and Gardner (1955) concluded that the New England foldbelt was the source of the rutile supplied to the adjacent younger basins. Gardner's conclusion, however, invoked an alteration of the abundant ilmenite to coarse single-crystal rutile, a process not yet observed in nature.

Whitworth (1959) showed that New England zircon and monazite were unsuitable source materials of the Quaternary coastal deposits, because the morphologies and compositions of these minerals differ in the two areas. Layton (1966) pointed out that cassiterite, a stable mineral supplied in the New England foldbelt, is lacking in most of the Quaternary coastal deposits. Whitworth was able to find very little rutile in stream sediments of the New England foldbelt; my own conclusions, based on stream sampling there, are emphatically in agreement with those of Whitworth.

If the New England rocks were not the ultimate source, what rocks were? Whitworth (1959) and Layton (1966) pointed out the suitability of some metamorphic rocks in the interior Australian craton. I would like to add that in Mesozoic time, Australia was still adjacent to other portions of the former Gondwanaland. Paleocurrent measurements from the Hawkesbury Sandstone (Rust and Jones, 1987) show transport to the northeast and make continents since removed just as suitable a source as the Australian craton (cf., Galloway, 1972).

History of development. The present sediment budget for rutile- and zircon-dominated shoreline deposits apparently contains a key to their development. These deposits are fed mainly by cannibalism of older but similar coastal deposits and by longshore drift from the south. The juvenile component in the mineralogy of this material must be small, as the supply by rivers is presently minor. The main source of the mere trickle of juvenile material must be sea-cliff and headland erosion, and this type of supply is far more productive of sand in the Sydney and Clarence-Moreton basins than in the New England and other foldbelts. Sea-cliff erosion of friable sandstones is characteristic of Mesozoic rocks of the basins. Thus the supply at present is small but is largely from sedimentary basins that supply almost exclusively rutile, zircon, ilmenite, tourmaline, and quartz to the system. Locally, headland erosion of other rocks dilutes the assemblage (Fig. 71).

Two lines of evidence suggest that the present sediment budget is not applicable during times of lower sea level. (1) Morphology and stratigraphy of river mouths suggest that at lower sea level, rivers become active suppliers of the system. (2) The high epidote and pyribole content of offshore sands of

[2]Ilmenite in shoreline deposits is likely to have a separate source in part, as some of it is too little weathered to have been cycled through Mesozoic sandstones.

both fluvial and shoreline origins is incompatible with derivation primarily from sandstones of the Sydney and Clarence-Moreton basins, suggesting that immature fluvial debris has swamped those sources. Sands offshore from the New England foldbelt are the debris of low-grade metamorphic and granitoid terranes, whereas those off the Sydney basin show the influence of basaltic intrusions[3] in the interior of the basin and of the adjacent Lachlan foldbelt (Hudson, 1986).

Lowered sea levels apparently activate a new distribution system, in which fluvial debris swamps debris from sea-cliff erosion (cf., Schluter, 1982). This sea-level "trigger" is undoubtedly modified, indeed intensified, by the coupled weathering–sea level factor described in Chapter 6. Sands now offshore that were formed in subaerial environments were subject to less maturation by weathering because temperatures were colder and atmospheres less reactive. In addition, the duration of in-place weathering was less because the sands were seldom exposed.

The changes in mineral distribution with sea level appear to explain (1) the great maturity and unique assemblage of all the interglacial-age shoreline sands, (2) the modified heavy-mineral assemblage of these sands adjacent to the New England foldbelt, and (3) the great contrast in both provenance and maturity of the sands offshore. Note that the mechanism proposed by Colwell (1982b), shoreward winnowing of offshore heavy minerals, is unnecessary and does not explain all the mineralogic data. The minerals onshore and offshore are related to each other not in terms of mechanical stability and density but in terms of differing weathering stability and provenance.

Geographe Bay (Bunbury-Capel) district, Western Australia

This district has been an important producer of ilmenite since 1956. Constituent deposits have been well described (Welch and others, 1975; Baxter, 1977), but the geologic context needs further study. Geologic maps of the area are by Lowrie and others (1967, 1983). The deposits lie as far north as Waroona (Fig. 74), but mining is centered on the Bunbury-Capel area. The district forms the southern end of the Swan coastal plain, facing the Indian Ocean. The area is farmed except for extensive forests of jarrah and other native trees. Relief is low, and rock exposure is poor, in part because of extensive eolian sand cover.

The heavy-mineral deposits are found in three former shoreline complexes ranging in age from early Pleistocene or Pliocene (Collins and others, 1986) to Holocene (Collins and Hamilton, 1986); each complex represents a progradational swarm of shorelines at progressively lower elevations. From oldest to youngest, these are the Yoganup complex, representing six individual shorelines at 66 to 26 m elevation; the Capel complex, with up to 10 shoreline deposits at 4 to 6 m elevation; and the Minninup com-

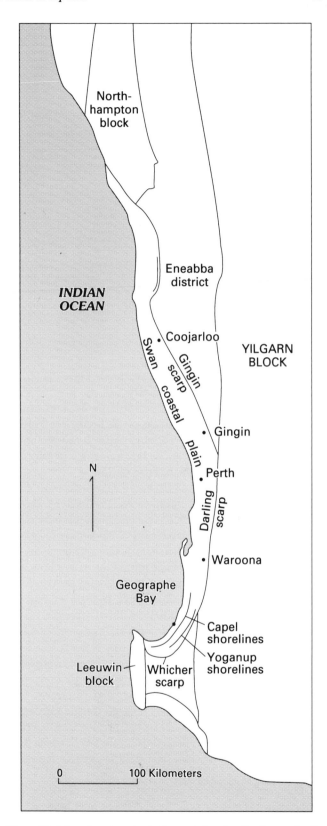

Figure 74. Map of western Australia coast showing the Geographe Bay and Eneabba districts, the Swan coastal plain, and various "scarps" and crustal blocks. The Minninup shorelines, not shown, coincide at this scale with the modern shoreline.

---

[3]My fluvial samples from the Sydney basin show a large contribution from basaltic rocks, probably the numerous diatremes of Crawford and others (1980).

E. R. Force

Figure 75. Waroona deposits at the foot of Darling scarp, western Australia. Here a former sea cliff coincided with an important structural feature bounding the Swan coastal plain. Portions of the deposit are on granitoid bedrock. The plain in the foreground is locally punctuated by former sea stacks.

plex, at or slightly above modern sea level. These shorelines outline successive paleo–Geographe Bays.

The Yoganup and Capel complexes are cut into Lower Cretaceous clastic sedimentary rocks of the Leederville Formation. A former sea cliff of the Yoganup system is exposed at the surface as the Whicher scarp and in the Waroona area coincides with the Darling structural scarp (Figs. 74, 75).

General description. All the heavy-mineral deposits of the district are dominated by ilmenite. As described by Baxter (1977), the deposits in the three shoreline complexes have other features in common also. They consist of moderately to poorly rounded grains in poorly sorted to bimodal progradational deposits. These are locally conglomeratic toward the base, with a matrix of heavy-mineral–rich clayey sand. Overlying the conglomerate is heavy-mineral–rich clayey sand, overlain in turn by fine-grained sediment. Interbedded local fluvial deposits are poor in heavy minerals. The surface deposits are transgressive eolian dune sands related in age to the next younger shoreline.

Heavy-mineral contents of greater than 10 percent are quite common in the deposits of this district. Ilmenite ranges from 56 to 95 percent of the heavy minerals, zircon 2 to 18 percent, and rutile 0.5 to 2 percent. Other heavy minerals include leucoxene, monazite, and kyanite; garnet, magnetite, pyriboles, and epidote are common in the younger deposits (Carroll, 1939), which contain calcareous shell fragments and fresh feldspar in the light-mineral fraction.

Differences among the shorelines in degree of weathering beneficiation are pronounced. The older shorelines have weathered mineral assemblages, including ilmenites having high TiO_2 contents (Fig. 52).

Geologic and depositional setting. Geographe Bay and its forerunners are developed in the Bunbury Trough, a graben filled

with sedimentary and minor volcanic rocks between the Leeuwin and southwestern Yilgarn crustal blocks (Fig. 74). Both blocks are composed of high-grade metamorphic and granitoid rocks. The two older shorelines were cut into sandstone of the Leederville Formation, which itself has a high heavy-mineral content in some places (Baxter, 1977, p. 47; Collins and Baxter, 1984). Its mineral assemblage is not recorded.

The modern shore of Geographe Bay is microtidal with sporadically high wave energy (Semeniuk and Johnson, 1982; Collins and Hamilton, 1986). The direction of longshore transport reverses seasonally. A typical succession of near-shore and beach depositional environments is present (Semeniuk and Johnson, 1982), with heavy-mineral concentration most efficient in the swash zone of the beach face (Hocking and others, 1982; Collins and Hamilton, 1986). Heavy minerals are transported by eolian action to transverse foredunes, where their concentration is less but their preservation potential greater (Baxter, 1977; Hocking and others, 1982; Collins and Hamilton, 1986). The stratigraphy of the younger Holocene deposits shows beach facies preserved by progradation (Semeniuk and Johnson, 1982).

Holocene sea-level history shows a general stillstand for about the past 7,000 years, locally modified by tectonic movements (Semeniuk and Searle, 1986; see also Welch and others, 1975, p. 1074). Progradation during this stillstand was preceded by transgressional landward displacement of shoreline sand bodies (Hocking and others, 1982).

Deposit facies and stratigraphy. Progradational stratigraphy of economic deposits is shown by Collins and Baxter (1984) and Collins and Hamilton (1986). Lithologies such as planar-laminated, fine-grained, well-sorted sand rich in heavy minerals are interpreted as high-energy beach deposits.

Progradation of beach deposits was apparently amplified by local fall of sea level, as progressively younger deposits formed on progressively lower wave-cut benches (Fig. 64B). Eolian sands having lesser heavy-mineral contents commonly overlie the beach sands and constitute the majority of some deposits (Welch and others, 1975).

Alteration and cementation. Alteration of the deposits of the Geographe Bay district has been documented as a function of elevation and thus of age (Fig. 52). The younger sands contain calcareous shell fragments, fresh feldspar, an immature heavy-mineral suite, and fresh ilmenite. The older sands are noncalcareous and contain a clay matrix partly from feldspar alteration and partly from infiltration by younger clays (Baxter, 1982). Heavy-mineral suites of the older sands are restricted to the more stable species, and ilmenite is altered, commonly concentrically, to contain about 60 percent TiO_2 (Welch, 1964; Baxter, 1977, 1986; Frost and others, 1983).

In the older deposits, degree of weathering is a function of position in the vertical weathering profile; the most altered assemblages are above less altered assemblages (Fig. 48). This weathering is clearly postdepositional (Baxter, 1982). So-called coffee rock, that is, sand cemented by iron hydroxides, is found along present and former water tables (Welch and others, 1975;

Baxter, 1982). The iron in this cement has been leached from overlying ilmenite grains, according to Welch (1964), who also suggested that iron cementation is triggered by the leaching of all pH-buffering calcareous material. Baxter (1986) showed that iron cementation arrests the alteration of encased ilmenite at an early stage.

Source. All authors seem agreed that the immediate source of ilmenite was the Leederville Formation; sea-cliff erosion of that heavy-mineral–rich unit must have provided abundant heavy minerals to beaches below. Southwestern Australia provides several possibilities for ultimate sources of this ilmenite. Baxter (1982) is inclined toward the southwestern Yilgarn block (Fig. 74) as the major source, whereas Carroll (1939) and Welch and others (1975) believe the Leeuwin block to be more important. Both areas contain high-grade metamorphic rocks (Wilson, 1964, 1969; Peers, 1975a).

Eneabba district, Western Australia

North of Perth is the Eneabba district (Fig. 74), the most important remaining identified resource of titanium oxide minerals in Australia except for the newly discovered Waroona deposits in Victoria. Related deposits south of the Eneabba district are near Jurien, Coojarloo, and Gingin. The Eneabba district was not discovered until 1970. It is valuable because of its great size, high heavy-mineral grades, and a rutile-rich mineral assemblage. Total heavy mineral contents of greater than 10 percent are common, and heavy-mineral resources are greater than 30 million metric tons. The district has been described by Lissiman and Oxenford (1973, 1975), Baxter (1977), and Shepherd (1990); except for attributed statements, these works are the basis for the following description. Geologic maps of the district are by Baxter (1972) and Lowrie and others (1973). Further study, especially sedimentological, is needed.

The Eneabba district is toward the northern end of the Swan coastal plain (Fig. 74). The greatest local relief is on the Gingin scarp, an aggregate of former sea cliffs cut into the Yarragadee Formation of Jurassic age. An eolian sand cover over thick laterite has produced poor exposures and a smoothed topography. Intensive farming to the south along the coast gives way within the district to sparsely settled dry scrubland to the north and east.

The major resource of the district consists of more than seven high-grade layers, each 1 or more meters thick. These layers were deposited with other clastics on wave-cut platforms forming subparallel steps from 130 m down to 29 m (Fig. 64C). These sediments, thought to be early or pre-Quaternary in age, are called the Yoganup Formation by Baxter (1982); that is, he correlates them with the Yoganup shoreline sands to the south. The sands are locally cemented by iron hydroxides and a clay matrix. Overlying and locally interbedded with these deposits are lower-grade eolian deposits. Shapes of eolian landforms suggest paleowinds from the southwest.

The modern Indian Ocean coastline of the district is carbonate dominated and normally of low to moderate wave energy.

Figure 76. Undercut portion of former sea cliff cut in the Yarragadee Formation, exhumed in mining of the zircon-rich 115-m shoreline, Eneabba, Western Australia.

Ambient winds are from the southwest, but monsoonal winds are from the northwest, producing variable longshore drift directions. Discussions of paleogeography follows.

The mineral assemblage of all the deposits is dominated by ilmenite, rutile, and zircon, with minor monazite, tourmaline, kyanite, staurolite, and pyribole. Average heavy-mineral grain size ranges from 0.15 to 0.18 mm.

Facies. The upper terrace deposits, from 128 m down to 100 m, represent beach deposition on benches cut in the Yarragadee Formation. Locally, the undercut surface of a sea cliff is exhumed by mining (Fig. 76). These deposits are high grade, and concentrates are rich in zircon (36 to 61 percent) and locally monazite, at the expense of ilmenite (28 to 46 percent) and rutile (5 to 8 percent). Minerals are well rounded and moderately sorted. Mean quartz grain size is 0.3 mm, and that for the bulk heavy-mineral fraction is 0.8 ϕ finer. Skewness of both light and heavy fractions is low. Deposition is apparently under conditions of high wave energy.

Deposits on lower terraces show several important differences. The heavy-mineral assemblage is dominated by ilmenite (53 to 68 percent) and rutile (8 to 11 percent), at the expense of zircon (15 to 23 percent). Lighter heavy minerals (staurolite, kyanite, tourmaline) are more common. Mean grain sizes of heavy minerals are similar to those on higher shorelines, and heavy minerals are well rounded and sorted. Average sizes of heavy minerals are related to each other as would be predicted from their densities. Light minerals, however, are reported to be finer grained than on the higher shorelines. Heavy minerals show appreciable positive skewness, whereas light minerals show strongly negative values. These data could be explained by the presence of two grain populations.

I was able to examine the 91- and 94-m shoreline deposits in a freshly excavated trench. Individual high-grade beds about 5 cm thick consist of rounded heavy minerals and quartz, supported partly by interstitial clay (Fig. 77). These beds show herringbone cross-lamination (Fig. 78), indicative of a low-energy intertidal environment. Separating the high-grade beds are interbeds of clayey sand virtually barren of heavy minerals. There is a suggestion of distortion of sedimentary structures by dissolution of carbonate shell fragments, but fresh feldspar is present. Such deposits probably formed on beaches of a coast with high tides but low wave energy. The two grain populations are a coarser rounded population generated by wave energy, probably higher on the beach and derived in part by sea-cliff erosion, and a finer population representing calmer intertidal conditions.

Overlying the shoreline deposits of the Eneabba district are (1) clayey estuarine sands, (2) deposits of alluvial fans off the Gingin scarp, and (3) several generations of eolian sands plastered across the Gingin scarp. The eolian sands are lower in heavy-mineral grade but are commonly mined with underlying shoreline deposits (Fig. 79); locally they form an independent resource. Lighter heavy minerals such as kyanite are relatively more abundant in the eolian sands. Under the eolian sands, a pisolitic laterite surface is commonly developed.

Alteration and cementation. The Eneabba shoreline deposits were intensely weathered, as shown by the lateritic surface developed on them. They now contain no carbonate, and some original feldspar has gone to granular clay lumps. Heavy mineral assemblages are mature, and ilmenite is altered to contain 62 percent TiO_2. Frost and others (1983) show that pseudorutile is the major constituent of this altered ilmenite.

The shoreline deposits are locally cemented by iron hydroxide, to form hard rock known locally as "coffee rock" that shows structures suggestive of nucleation around tree rootlets and/or marine bioturbation (Fig. 80). It probably represents precipitation, at former water tables, of iron derived in part from leaching of ilmenite in overlying sands. Ilmenite encased in coffee rock is less altered than that in permeable sand (Lissiman and Oxenford,

Figure 78. Herringbone cross-lamination in deposits of the 91-m shoreline, Eneabba. Lens cap 48 mm in diameter.

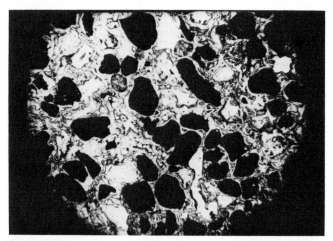

Figure 77. Photomicrograph of impregnated ore from the 91-m shoreline, Eneabba. Note dark clay matrix. Detrital minerals are ilmenite, rutile (both dark), zircon, and quartz. Transmitted plane light, 2-mm field.

1975). Cementation apparently started early, as detrital fragments of coffee rock are incorporated in the youngest Eneabba shoreline deposits.

Source. Jurassic rocks of the Cockleshell Gully, Yarragadee, and South Perth formations are important intermediate sedimentary sources of the titanium minerals of the Eneabba district. Local high heavy-mineral contents are reported from the Yarragadee and South Perth formations; friable heavy-mineral–rich sandstones of these formations may themselves constitute low-grade resources. My collection of Yarragadee sandstones from throughout the area suggests that the ratio of zircon to ilmenite and rutile varies considerably, from about 0.06 to 0.15. The low value occurs in a micaceous sandstone with relatively high contents of the lighter heavies staurolite, kyanite, tourmaline, epidote, and pyroxene.

The ultimate sources of the minerals of the Eneabba district are the Northhampton block to the north and the southwestern Yilgarn block to the south. Both contain abundant high-grade metamorphic rocks (Wilson, 1969; Peers, 1975b). A clue to ultimate sources is provided by rutilated blue-gray quartz in both the Eneabba deposits and in the Yarragadee Formation; this type of

Figure 79. Stratigraphy of Eneabba deposits in a mine face. Light-colored basal deposits are the upper surface of Mesozoic sandstones, into which heavy minerals have been introduced in joints and potholes. The dark overlying material is a heavy-mineral–rich shoreline deposit. Not visible is a laterized upper surface on this material. The uppermost light-colored material is eolian sand.

quartz is characteristic of granulite terranes (see Herz and Force, 1987, for details). Welch and others (1975) showed that the trace-element contents of Eneabba ilmenites are quite different from Geographe Bay ilmenites, implying separate sources for the two districts.

History of development. Lissiman and Oxenford (1973, 1975) and Baxter (1977) discuss the paleogeography of the district in terms of a north-facing former bay, the western side of which is a peninsula ("Rocky Springs cape") formed of Cockleshell Gully Formation, stratigraphically below the Yarragadee. This cape, with a maximum present elevation of 95 m, would have sheltered the depositional environments of the lower shorelines and would have added the Cockleshell Gully as a source for lower shorelines (Fig. 81).

Shepherd (1990), reporting on more recent exploration, finds no evidence for the Rocky Springs cape or the north-facing bay and no subcrops of Cockleshell Gully Formation in the mine area. In Shepherd's view, the different character of the higher shoreline deposits is due to selective preservation; only high-energy swash zone deposits were preserved on the narrow higher terraces.

Travancore Coast district, Kerala and Tamil Nadu States, India

This district, mined for monazite since 1911, became the world's largest ilmenite producer in the 1940s. Extensive use of hand labor and shallow harbors have limited production and caused the district to lose its position.

The geology of the Travancore deposits has been studied most comprehensively by Tipper (1914). Other studies are by Gillson (1959) and Mallik and others (1987). Other, somewhat

similar deposits of the Indian subcontinent are those in Orissa and northeastern Sri Lanka.

Two main deposits form the district, the larger deposit north of Quilon, on the Kayankulam-Needakara (K-N) bar of the Kerala coast, and the other at Manavalakurichi (MK) in Tamil Nadu (formerly Madras), about 100 km to the south, virtually at the southern tip of India. The borders of the district are exceedingly diffuse; many smaller coastal deposits are known, but in those more distant, ilmenite concentrates have low TiO_2 contents (Jacob, 1956; Sinha, 1967).

The MK deposit is in an embayment bounded by headlands of deeply weathered crystalline rock of the region (Gillson, 1959). The coast has been uplifted so that former sea cliffs are now elevated and inland. The K-N bar of the Quilon deposit is a barrier island on a drowned coast. At the southern end are headlands 2 to 10 m high of weathered Cenozoic Warkalli sedimentary rocks (Poulose, 1972; Prakash and Verghese, 1987). No fluvial sediment currently reaches the sea near the Quilon deposit (Tipper, 1914; Gillson, 1959), and net erosion occurs along this shoreline (Prakash and Verghese, 1987).

Both deposits are apparently Holocene in age and consist of high-grade enrichments on modern beaches, buried progradational beaches, and lower-grade eolian dunes (Gillson, 1959). Enrichment is only in sands at or above present sea level. Ilmenite is the predominant heavy mineral in both deposits (70 to 80 percent [%], with lesser zircon (about 5%), rutile (4%), sillimanite (4%), and monazite (0.5 to >1%). Garnet is common at MK but not at Quilon. Ilmenite concentrates from Quilon contain 60 to 62 percent TiO_2, whereas those from MK contain 54 percent (Sinha, 1967). The alteration of MK ilmenite has been described by Subrahmanyam and others (1982), and of Quilon ilmenite by Karkhanavala and others (1959). Rutile concentrates are reported to show high niobium and tantalum contents (Subrahmanyam and Rao, 1980).

Mining on modern beaches of the district has shown the formation of natural concentrates to be highly seasonal. Black

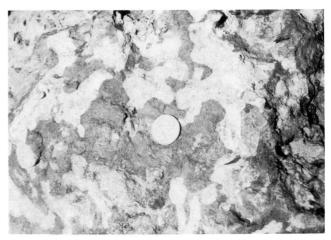

Figure 80. "Coffee rock" from Eneabba, containing iron hydroxide cement probably pseudomorphic after organic structures.

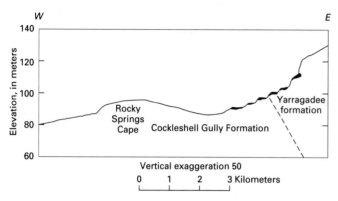

Figure 81. Cross-sectional relation implied by Lissiman and Oxenford (1973, 1975) and Baxter (1977) for shorelines, Rocky Springs cape, and exposed Mesozoic sandstones in the Eneabba district. See text for evaluation.

sands carpet the swash zones during strong onshore winds of the monsoon season, whereas otherwise the beaches are white and no concentrates are visible (Canadian Mining Journal, 1956). Concentration efficiency is so great that glass sand deposits are present also in modern shoreline sands of the same district (Poulose, 1972). In the Quilon sector, mining activity is gradually depleting the heavy-mineral stocks available for concentration (Gillson, 1959).

The ultimate sources of the heavy minerals of the district are granulite-facies gneisses of the interior. The most abundant lithologies of the interior headwaters of the district are khondalites (Balasundaram, 1970; Mallik and others, 1987); that is, sillimanite-garnet-graphite gneisses that here contain ilmenite and monazite (Soman, 1985) and commonly elsewhere contain rutile (Force, 1976b; Fig. 4).

Three intermediate sources are of importance in the district. First, thick laterites are developed on the crystalline rocks, and these release the right mineral suite (Gillson, 1959). Second, the Cenozoic Warkalli sandstones contain heavy mineral concentrations and are themselves weathered (Tipper, 1914). Poulose (1972) observed that the area of abundant sea-cliff erosion of Warkalli sandstones is precisely the area of rich Holocene heavy-mineral concentrates. Third, Pleistocene cemented coastal eolian dunes are being eroded (Tipper, 1914; Narayanaswami and Mahadevan, 1964) and their contents resupplied to beaches.

A significant role of weathering in the formation of sands of the district is implied by the beneficiated mineral assemblages and the numerous intermediate hosts (Poulose, 1972). Garnet, for example, is abundant in the source rocks and is present in the Warkalli beds but is virtually absent in the Quilon deposit (Tipper, 1914). Scanning electron microscope photographs by Mallik (1986) show a great deal of chemical etching that predates final rounding (e.g., his Figs. 2A, 3B).

Gillson (1959) proposed a Cenozoic history of stream capture in the fluvial drainages of the district. His model was appar-

ently the first to link source rocks and changes in fluvial conduits and sea level with the locations of any shoreline placer deposits.

Richards Bay district, South Africa

Heavy-mineral deposits north of Richards Bay, mined since 1967, have put South Africa in the forefront of titanium-mineral–producing countries. An innovative smelting process for the relatively low-TiO_2 ilmenite is partly responsible for the commercial success of the venture. Richards Bay is on the Natal coast, facing the Indian Ocean. The deposits have been described only by Hammarbeck (1976) and Fockema (1986). Quaternary development of coastal eolian landforms has been described by Maud (1968).

The economic heavy-mineral deposits consist of eolian sands in a belt along the modern coast, up to about 1 km inland and 5 to 25 km north of Richards Bay. They are apparently mostly Holocene, but some of the older sands that form part of the deposit may be Pleistocene, based on their weathering and cementation and the dating of buried stone implements (Davies, 1970). The eolian dunes are locally more than 100 m high and rest on a platform of 20 to 40 m elevation, underlain by the Pleistocene Port Durnford Formation. Some laterite is developed on this platform. The eolian beds rest on a beach sand that Fockema (1986) includes in the Port Durnford, but which Maud (1968) and Hammarbeck (1976) regard as post–Port Durnford. The upper part of the Port Durnford Formation is exposed in wave-cut benches of the district and, according to Maud and Hammarbeck, consists of yet more eolian sands.

The dune deposits are commonly about 20 m thick and average 10 to 14 percent heavy minerals. Heavy-mineral content is greater in the lower dunes and in the lower parts of the high dunes. The economic heavy minerals ilmenite, zircon, and rutile average 5.9 percent of bulk ore. Ilmenite is the dominant heavy mineral and contains 46 to 50 percent TiO_2. Some leucoxene and monazite are present and are also of value. Other minerals present include hornblende and augite, magnetite, garnet, epidote, and tourmaline. Feldspar is present in the younger sands. Average diameter is about 0.3 mm for the light minerals and about 0.1 to 0.15 mm for the heavy minerals.

The chain of probable intermediate sources is long. The Port Durnford beds are clearly important sources of the heavy minerals in the dunes. Upper Cretaceous sandstones also crop out along this coast. Hammarbeck (1976) believes that the Ecca Series (Permian) of the Karoo System is an intermediate host of some importance, based on similar ilmenite composition. Karoo rocks are the most abundant within a 200-km radius of the district, and elsewhere in the Karoo basin contain Permian shoreline-facies concentrations of the same mineral suite (Behr, 1965, 1986). The presence of little-altered ilmenite with magnetite, pyroxene, and epidote in a mineral suite including leucoxene (Dimanche, 1972) suggests multiple intermediate sources for heavy-mineral sands of the district. The ultimate sources are probably granulite-facies gneisses, some of which are present in the immediate area (Saggerson and Turner, 1978).

ECONOMIC PROGNOSIS

Quaternary shoreline deposits are currently the most important source of titanium minerals, and for most uses they are the preferred source. The only problem with the continued commercial success of shoreline deposits relates to their possible exhaustion. Deposits of this class will probably continue to capture a respectable share of the market as long as sizable deposits remain accessible.

Though many currently known shoreline placers have limited lives, this does not mean that the future of the entire deposit class is sealed. As this is written, two large, newly discovered shoreline placers, in Victoria, Australia, and Madagascar, are being developed, reflecting the fact that much remains to be learned and done. Exploration efforts to date have used a small fraction of the pertinent information.

In some areas of the world, environmental constraints have prevented or halted exploitation of shoreline placer deposits that are still forming or are of very recent age. This is legitimate, and we should expect environmental protection of modern coastlines to become more common worldwide. In most districts, protection of modern coastlines takes away only a small fraction of the resources there. In the same area where the modern shoreline is subject to high land values, competing land uses, and strict environmental regulations, multiple deposits of former shorelines a few kilometers inland are typically free of these constraints. The only advantage of the modern over the former shoreline is in ease of discovery; the potential for deposits is the same, and the amount of exploration required to confirm a deposit is the same. The older deposits commonly have as little overburden as the modern deposits and are more beneficiated by weathering.

METHODS OF EXPLORATION

Conceptual exploration

Every portion of this book, and probably much more information besides, bears on the exploration for shoreline placer deposits of titanium minerals. These deposits can form where (1) source rocks have the proper mineralogy, (2) weathering restricts the heavy-mineral assemblage, and (3) conduits can supply the right material in little-diluted form. Only then can concentration on the beach face form an economic deposit. Given these prerequisites, additional factors may aid the concentration process and the preservation of the concentrate. All these factors can be considered *in sequence* in exploration programs to delineate promising areas.

Source areas are logically considered first, as they are necessary to supply the proper mineral assemblage. The great majority of economic shoreline placer deposits of titanium minerals clearly have ultimate source areas of high-grade metamorphic rocks (Force, 1976b; Goldsmith and Force, 1978). These rocks contain the proper mineral suites (Chapter 2) and form large source terranes (Force, 1980b). Anorthosite-ferrodiorite massifs commonly found with these terranes also supply the proper mineral suite.

Intermediate sedimentary sources are of great importance in every district described. In the eastern Australia district, such intermediate sources are the only ones we see, possibly because of continental fragmentation. The exploration logic relative to source should be to consider coasts currently fed by favorable ultimate source terranes and/or by clastic sediments and sedimentary rocks known to have favorable mineralogy.

Weathering is the next permissive condition to be considered. Postdepositional weathering is known to be important, but every district described in this chapter shows strong evidence of predepositional weathering at other sites. Among these sites, the intermediate sedimentary hosts are probably most important. Simple zonation by latitudes apparently controls the distribution of currently economic Quaternary shoreline placer deposits, with auxiliary influence by rainfall distribution. However, changes in climate through time clearly influence intermediate hosts and pre-Quaternary placers. The working of these factors is detailed in Chapter 6. The exploration logic relative to weathering should be to look for Quaternary shoreline deposits on coasts of moderate relief at latitudes lower than 35°, unless climate change is thought to have produced weathering-beneficiated assemblages in intermediate hosts. For Quaternary shoreline deposits now offshore, the favorable latitudes are restricted as a function of depth (Chapter 6).

Conduits that supply the proper detrital feed to the shoreline in little-diluted form are the next consideration. Fluvial systems that characteristically supply little-diluted material are described in Chapter 8. In this chapter, however, we have seen that conduits integral to the coast itself are equally important. These commonly involve sea-cliff erosion of favorable intermediate hosts and/or the suppression of fluvial or other unfavorable debris by sea-level change. The most valuable deposits probably result where fluvial drainages bring favorable weathered material to the coast in a little-diluted state and/or where coastal processes at particular sea levels bring material from favorable intermediate hosts.

Guild (1971) stated that shoreline placer deposits of titanium oxide minerals occur on passive continental margins. Statistically he is right. Such margins permit favorable source rocks to be adjacent to the coast, permit deep weathering because of moderate relief, and permit delivery to the coast of fluvial mineral suites little contaminated by volcanics, glaucophane, serpentine, and other troublesome components of active continental margins. However, Guild's concept neglects the possibility of microplate accretion of favorable terranes onto an active margin. The coastal anorthosite-granulite terrane of Oaxaca, Mexico, is an example. Somewhere else on active margins a sedimentary intermediate host bearing weathered titanium oxide minerals could have been rafted in to form coastal outcrops; fluvial debris from the remainder of the margin could be suppressed by sea-level rise, and the valuable titanium minerals supplied only to local coastal compartments.

Favorable depositional environments

The next step is the search for physical environments that produced and preserved placer concentrations of titanium minerals. Progradational high-energy beach systems are a promising target; storm and seasonal events produce concentrations on swash faces, and burial by progradation preserves them. Enrichment in such deposits commonly occurs in the upper portions of beach-ridge sand bodies and represents the portion of the beach complex that accumulated at and above sea level.

Transported enrichments may occur in eolian sand masses. These are derived from the coast but may or may not be physically contiguous with true beach deposits; some coastal eolian sands are found atop sea cliffs. Assuming no overburden, these sand masses will be topographically positive elements, commonly but not always elongate parallel to the paleoshore. Heavy-mineral concentration in eolian deposits is likely to be fairly homogeneous, except at the scale of individual laminae. However, the tops of the highest dunes are reported to be impoverished in heavy minerals in some districts.

The positive geomorphic expressions of former barrier islands and associated foredunes have been widely used in exploration for shoreline placer deposits. Geomorphology may indeed give some strong clues to the locations of different depositional environments and even conduit systems. Lithologic and/or soils maps, if available, can indicate the locations of appropriate sand bodies, and used in conjunction with geomorphology, can efficiently direct exploration. Air-photo interpretation may help delineate progradational beach-ridge systems and transgressive dune morphology.

In many districts, certain parts of individual shoreline compartments or barrier segments are consistently those showing greater heavy-mineral concentration (cf., Peterson and others, 1985). These are apparently controlled by the locations of headlands and/or inlets. Once the pattern has been established in a district, geomorphic analogs can be systematically prospected.

Geophysical exploration

Exploration based on the physical properties of shoreline placer concentrations has been instrumental in the discovery and delineation of several deposits, and there is great potential for further improvement of these methods. Methods investigated thus far include magnetism, radioactivity, and induced polarization.

Economic shoreline deposits of titanium oxide show only very subtle magnetic response (Robson and Sampath, 1977) because (1) these bodies are shallow, (2) favorable source terranes contribute little magnetite, and (3) the weathering required for beneficiation of the mineral assemblages removes what magnetite is present.

Aeroradioactivity has been the most successful geophysical exploration method thus far, and large portions of some coastal plains have been thus surveyed. The Altamaha Plantations and Green Cove Springs deposits were delineated in part from total-count gamma aeroradioactivity maps. Detailed studies (Force and others, 1982; Grosz, 1983) have shown that a great deal of geologic knowledge is necessary for correct interpretation of aeroradioactivity maps and intelligent exploration with them. Surficial heavy-mineral concentrations can form aeroradioactivity highs where monazite and zircon are present. However, these highs may be hidden by thin overburden, including water, or they may be swamped by other common surficial lithologies that produce high radioactivity values. In addition, human activities and additions to the surficial environment—for example, phosphate fertilizer—may greatly modify the aeroradioactivity pattern (Force and others, 1982; Grosz, 1983). Spectral radioactivity surveys, whether air- or ground-based, can help sharpen exploration efforts with aeroradioactivity, as heavy-mineral concentrations are normally dominated by the contribution from thorium. Aeroradioactivity prospecting coupled with sampling has shown that most shoreline sands have radioactivity responses *lower* than ambient values because surficial exposures of most shoreline sands are poor in heavy minerals (Force and others, 1982). Thus, surface exposures of shoreline sands rich in heavy minerals commonly produce local positive aeroradioactivity anomalies in larger areas of low values.

Ground-based induced polarization (IP) methods have proved to respond to shoreline placer deposits dominated by altered ilmenite (Wynn and others, 1985) but not to those dominated by rutile (Robson and Sampath, 1977). Spectral IP surveys appear to differentiate between the response of ilmenite and other minerals.

Geophysical methods of exploring for deposits now offshore require modification. Radioactivity exploration is hampered because gamma radioactivity is shielded by water. Induced polarization, on the other hand, is simpler than on land, because the conductivity of sea water permits use of a surface-towed streamer (Wynn and Grosz, 1986).

Chapter 10.

Placer deposits in pre-Quaternary sedimentary rocks

Placer deposits described in preceding chapters are those in which original depositional landforms are still preserved. Some possibly pre-Quaternary deposits that met this criterion were included. The pre-Quaternary deposits described in this chapter are buried by younger sediments, thus masking original depositional landforms. Some such deposits even have steep dips. In two districts described, host rocks are indurated, whereas in two others they are not. The inherent economic disadvantage of an indurated deposit can be overcome only with very high grades. Ironically, some details of sedimentary deposition are best seen in cross section in the indurated deposits.

Only one pre-Quaternary placer district of titanium oxide minerals, in the Lakehurst area of New Jersey, has been mined extensively. This deposit is closely analogous to Quaternary placer deposits in many respects. The economic status of the other pre-Quaternary deposits is uncertain.

DEPOSITIONAL ENVIRONMENTS AND TEMPORAL VARIATION

All pre-Quaternary placer concentrations of titanium minerals described to date as economic or near economic are shoreline deposits and thus formed under conditions described in Chapter 9. Stratigraphic relations detailed in this chapter show that the deposits formed on prograding shorelines, that is, they are underlain by marine rocks and overlain by nonmarine rocks. Individual concentrations probably represent minor erosional transgression, like concentrations observed in modern prograding beach systems. In two districts described here, stratigraphic descriptions are sufficiently precise that the environment of concentration can be specified as the upper swash zone. In one district, the distribution of heavy-mineral concentrations relative to other lithologies has been used to hypothesize details of the paleoenvironment that would be difficult to reconstruct in any other way (Houston and Murphy, 1962, 1977).

Eolian deposits have not been preserved in any of the districts described in this chapter. If eolian deposits were originally present, they probably were planed off in the nonmarine environments represented by the beds overlying the beach concentrations, as the dunes were mobile positive elements of the land surface. The prevalence of titanium-mineral resources in Quaternary coastal eolian deposits probably reflects lack of reworking during burial.

Three of the four districts described in this chapter formed during geologic time intervals characterized by deep weathering (Late Cretaceous, Miocene-Pliocene boundary). Their mineral assemblages are thus more beneficiated by weathering. The importance of such weathering, and temporal variations thereof, are discussed in Chapter 6.

INTRASTRATAL SOLUTION

The leaching of heavy-mineral assemblages, discussed in Chapter 6 in terms of surficial weathering processes, acquires a subsurface dimension in buried older deposits. The entire heavy-mineral assemblage of a sandstone can vary greatly with its history of intrastratal solution. Equilibration with a variety of ground waters may have occurred over relatively long time periods. An extensive summary is given in Pettijohn (1957). Iron silicate minerals are particularly susceptible to alteration, both in oxidizing (Walker, 1967) and reducing (Siever and Woodford, 1979) ground waters. Skeletal remnants of these former detrital minerals are visible in some sedimentary rocks (Houston and Murphy, 1962; Friis, 1974; Scholle, 1979). The titanium silicate sphene is also somewhat unstable and may be replaced by "leucoxene" and quartz and/or carbonate (Morad and Aldahan, 1985).

Detrital ilmenite may alter to authigenic anatase, rutile, and/or brookite, forming rims, crack fillings, and/or cement (Houston and Murphy, 1962; Mader, 1980; Morad and Aldahan,

1986). Authigenic anatase is locally euhedral and too coarse to be considered microcrystalline, but it is finer than associated detrital sand grains (Houston and Murphy, 1962; Morad, 1986). Rutile as skeletal aggregates may form pseudomorphs of detrital ilmenite where alteration of ilmenite is complete. Iron sulfide minerals, where present with titanium dioxide alteration products, suggest that leaching of iron was under reducing conditions (Dimanche and Bartholome, 1976; Reynolds and Goldhaber, 1978).

MAJOR DEPOSITS

Lakehurst district, New Jersey

From 1962 to 1982, the Lakehurst district was an important source of altered ilmenite in the United States. Mining, by two companies in separate pits, was largely from the upper Tertiary Cohansey Sand.

The district is in the "pine-barren" region of the coastal plain of southern New Jersey. Population growth in the area is rapid, and part of the district lies beneath a naval air station used for lighter-than-air craft. Topographic relief is low; average elevation of the land surface at the deposits is about 30 m. The district was probably continuous originally, but fluvial dissection during the Pleistocene and Holocene has produced separate deposits.

Ilmenite deposits of the Lakehurst district were discovered in 1956 by Frank Markewicz and his colleagues at the New Jersey Geological Survey, using a genetic model involving source rock and fluvial conduit (Markewicz, 1969). The most substantial of the more recent studies are by Carter (1978), who documented stratigraphic trends in depositional environments, and Puffer and Cousminer (1982), who presented an integrated sedimentologic-paleoclimatic-mineralogic analysis of the deposit.

General description and stratigraphy. The ilmenite deposits are overlain by up to 7 m of Pleistocene(?) fluvial gravels (Quirk and Eilertson, 1963). The upper Tertiary formations hosting the deposits are unindurated. The highest-grade portions, which contain greater than 5 percent heavy minerals, form about 5 m of laminated fine to medium sand near the middle of the Cohansey Sand, which here is about 20 m thick (Quirk and Eilertson, 1963; Puffer and Cousminer, 1982). Altered ilmenite is the predominant heavy mineral (85 percent); also present are zircon (7 percent), sillimanite (3 percent), and staurolite and tourmaline (each 1 percent). Underlying the Cohansey are finer sands of the Kirkwood Formation that contain lesser concentrations of the same heavy-mineral assemblage. Recent workers are inclined to regard the base of the Cohansey as conformable with the underlying Kirkwood.

Carter (1978) has divided the Cohansey Sand into two "sequences," with the heavy-mineral–enriched zone in the lower sequence. Shallow-marine trace fossils are found throughout the formation. At the base of the Cohansey are interbedded granule and sand layers that represent deposition in the surf zone. An overlying laminated facies is enriched in heavy minerals and represents swash-zone deposition. The lamination dips seaward, that is, toward the south-southeast. Overlying this facies in an area of about 1 km^2 is the Legler Lignite (Rachele, 1976), a freshwater swamp deposit 1 to 5 m thick with thin associated clays that contain marine microfossils in the upper part of the unit. These lithologies, forming Carter's lower sequence, clearly show beach progradation. The upper sequence of the Cohansey contains cross-bedded subtidal and intertidal sands, and in my opinion represents a separate depositional sequence following transgression.

Environment of concentration. Puffer and Cousminer (1982) found that sands enriched in heavy minerals (as much as 63 percent) are fine (0.2 mm mean) and well sorted, but not well rounded. They attributed high positive skewness to an eolian origin but neglected to consider that heavy-mineral enrichment in shoreline environments occurs by preferential removal of coarse lights, which necessarily produces positive skewness. The stratigraphic position of greatest enrichment recorded by Carter (1978) was at the top of the swash-zone facies, and this suggests upper swash-zone concentration similar to modern beach deposits.

Ilmenite alteration. Mathis and Sclar (1980) and Puffer and Cousminer (1982) found that precursor ilmenite-hematite intergrowths were pseudomorphically replaced by microcrystalline pseudorutile (topotactic after ilmenite) and voids (after hematite lamellae). This alteration is less complete in the Kirkwood than in the Cohansey, where an appreciable number of grains also show microcrystalline rutile. In the Cohansey, altered ilmenite averages 65 percent TiO_2. Unable to relate weathering state to depth below land surface, Puffer and Cousminer suggested that an appreciable component of ilmenite alteration was predepositional. Some postdepositional alteration, however, is suggested by local iron hydroxide cement, which encases minor magnetite as well as ilmenite (James Stern, personal communication, 1974). The intensity of alteration, coupled with the subtropical nature of Legler Lignite palynomorphs, prompted Puffer and Cousminer to hypothesize a warmer climate during deposition than at present.

Geologic evolution. Puffer and Cousminer (1982) envision the Lakehurst deposits derived from ilmenite-rich gneissic source rocks in the Hudson Highlands and possibly the Adirondack Mountains and from older sedimentary hosts, during a period of deep weathering at the boundary between the Miocene and Pliocene epochs. Deposition was during a sea-level regression of that period, followed by marine transgression. The ancestral Delaware River was the fluvial conduit, as suggested by Markewicz (1969).

McNairy Sand, Tennessee

The Upper Cretaceous McNairy Sand contains substantial resources of heavy minerals in three deposits that define a district in western Tennessee (Wilcox, 1971). The McNairy is an unindurated, very fine-grained sand representing shoreline deposition in the Mississippi embayment from Mississippi to Illinois. The heavy-mineral assemblage is economically attractive, and ilmenite is altered to TiO_2 contents of about 60 percent, but fine grain size

Figure 82. Heavy minerals in facies of McNairy Sand exposed in a silica sand pit operated by the Jessie Morie Co. in the Seventeen Creek quadrangle, Tennessee. A. Flat-laminated facies below, with heavy-mineral concentration above shovel, overlain by lowest cross-bedded facies. B. Large cross beds outlined by heavy-mineral laminae in the upper unit.

makes mineral separation difficult. The district is not currently being developed.

Physiographically, the McNairy Sand forms a dissected low plateau bordering the presently dammed Tennessee River. The forested Natchez Trace State Park occupies the southern end of the district and contains substantial resources (Hershey, 1966, 1968). The McNairy is at the surface over much of the district but is poorly exposed except in steep bluffs along digitate margins of the plateau. The best exposures are in several silica sand mines in the McNairy.

The regional geologic context of the district has been described by Russell (1975). Quadrangles containing deposits have been mapped by Hershey (1966, 1968), Russell (1967), and Ferguson and Garman (1970). The McNairy Sand, about 50 to 100 m thick, is the shoreline facies of the upper regressive sequence in a transgressive-regressive wedge. The McNairy is underlain by a shallow-marine shale. Toward the eastern (landward) edge of the outcrop belt, at the closure of the wedge, the McNairy locally lies directly on Paleozoic bedrock. Northward into Kentucky, I have found that the McNairy becomes a tidal-flat facies with tidal channel sands, probably reflecting the stronger influence of delta growth to the north, as reported by Pryor (1960). Heavy-mineral contents are low in the McNairy of Kentucky, as they are in southern Illinois (Hunter, 1968).

The basal member of the McNairy Sand, up to 15 m thick, contains most concentrations of heavy minerals. The sands of this member show evidence of beach deposition. The very fine angular sands of the basal member locally contain more than 5 percent

heavy minerals, with an average heavy-mineral grain size of about 0.1 to 0.06 mm throughout the district. The average relative abundance of heavy minerals is altered ilmenite, 55 percent (%); leucoxene, 8%; rutile, 2%; zircon, 10%; monazite, 1%; aluminosilicate minerals, about 20%; and tourmaline, 2%. Ilmenite, zircon, and monazite are concentrated in the finest fractions (<0.06 mm), whereas less-dense leucoxene, aluminosilicates, and tourmaline are concentrated in the coarser (>0.06 mm) fraction. Wilcox (1971) pointed out that the relative grain size of leucoxene implies that it was already less dense at deposition and hence was altered before deposition. In two of the deposits, leucoxene is more abundant toward the base of heavy-mineral enrichments (Ferguson and Garman, 1970; "Manleyville" of Wilcox, 1971), also suggesting predepositional alteration.

In several silica sand pits southeast of Bruceton, the stratigraphic sequence in and above the basal member of the McNairy can be seen (Fig. 82A). Heavy mineral enrichments occur in fine- to medium-grained, well-sorted but angular sand with planar laminations dipping very gently southwest. *Ophiomorpha* burrows are common in this interval. These beds are apparently beach deposits. Overlying sands show increasingly high-angle cross beds with variable mica and clay contents and clay beds draped over dune and channel bed forms (Fig. 82B). These beds are apparently fluvial in origin and suggest sedimentary transport to the southwest. They contain fewer heavy minerals than the beach deposits, and their mineral assemblage contains far greater proportions of the lighter heavies leucoxene, kyanite, and goethite. The upper beds are preferentially mined for silica.

Upper Cretaceous deposits of the western interior, United States

Heavy-mineral concentrations in sandstones of Late Cretaceous age are known from Montana, Wyoming, Utah, Colorado, New Mexico, and Arizona. Several of these sandstones are black from heavy-mineral concentrations; most were found as a result of their radioactivity. The concentrations are preferentially cemented and therefore are resistant to erosion; they form hogbacks and mesas in several individual sedimentary basins. Sedimentologic and tonnage aspects of the deposits have been described by Houston and Murphy (1977) and Dow and Batty (1961), respectively. Houston and Murphy (1962) and Roehler (1989) have described the Wyoming deposits in some detail. These references are the sources of the following descriptions. All the deposits from six states are described here as a single district, for want of space for more detailed treatment.

The deposits represent regressive shorelines from several transgressive-regressive cycles in the Upper Cretaceous of the western interior seaway. Heavy-mineral contents are locally as high as 90 percent, and intervals as thick as 3 to 5 m may contain 10 percent or more heavy minerals. Opaque oxide minerals dominate the heavy-mineral assemblage throughout the district but vary from mostly magnetite in Montana to mostly altered ilmenite (with up to.60 percent TiO_2) in New Mexico (see also Chenowith, 1957). In the Wyoming deposits, altered hemoilmenite predominates over altered ilmenite. Zircon is common, and minor garnet, tourmaline, rutile, and monazite are present. In sandstones containing leaner heavy-mineral concentrations, garnet, tourmaline, sphene, amphibole, epidote, and staurolite form an appreciable part of the assemblage. The deposits are cemented by goethite, carbonate, and authigenic anatase (locally brookite). The goethite component of this cement apparently results from alteration of opaque minerals in the modern weathering environment.

Most of the deposits are in sequences of fine sandstone, but the rich concentrations are generally very fine sands. Constituent grains are mostly well rounded, although opaque oxide minerals are commonly rimmed by authigenic anatase, which gives them a variety of shapes.

Horizons showing the greatest concentrations are toward the top of stratigraphic intervals that represent beach deposition. These are underlain by shallow-marine deposits and are overlain in turn by lagoonal, swamp, and/or fluvial deposits. Enriched horizons and underlying beach sands contain planar laminae dipping at very low angles down Cretaceous beach faces (see cover illustration). Thus the concentrated horizons apparently represent prograding upper swash zone sands. Eolian sands have only been recognized in one locality (Roehler, 1989).

The elongation of these beach deposits in map view has been used to establish Cretaceous shoreline trends. Locally, Houston and Murphy (1962, 1977) have been able to hypothesize wind directions, tidal ranges, and other subtle features of the depositional environment based on variations in deposit geometry.

Karoo Basin, South Africa

Heavy mineral concentrations of economic interest occur in shoreline deposits of the Ecca Group, of Permian age, in the northern Karoo Basin. The deposits, though rich, are not economic because of difficulties with mineral separation. The thorough description of the Bothaville deposit by Behr (1965) is the basis of the following discussion. Behr (1986) has described other deposits of the district. The petrography of heavy-mineral alteration has been treated by Mukerjee (1964).

Enrichments occur in the Vryheid Formation (Middle Ecca sandstone), which rests on upper Paleozoic glacial deposits and Precambrian lavas. A swarm of individual deposits, each oriented northwest-southeast and dipping very gently southwest, occurs in progradational relation. Individual deposits average 3 m thick and 350 m wide; one is about 5 km long. The general stratigraphy of the Ecca Group suggests deltaic progradation, and heavy-mineral enrichments are thought to represent shoreline reworking of inactive delta lobes. Eolian facies are apparently absent.

Sands enriched in heavy minerals are well sorted and positively skewed, with average grain size 0.1 to 0.2 mm, in contrast to less enriched adjacent sandstones that are more poorly sorted and have average grain sizes of 0.25 to 0.30 mm. Detrital grains are well rounded.

Heavy-mineral contents of more than 50 percent are common in enriched beds. Ilmenite averages 60 percent, zircon 7 percent, and garnet, monazite, and rutile a few percent each, of the heavy-mineral suite. Some chromite is present. Ilmenite, which commonly contains hematite intergrowths, is rimmed by leucoxene consisting of microcrystalline anatase. Quartz, feldspar, and mica are the light minerals. Iron sulfides, siderite, chlorite, and anatase are present as cement; near the ground surface, goethite is present in place of sulfides and siderite (Mukerjee, 1964).

Vos and Hobday (1977) concluded from stratigraphic trends within the beach deposits that heavy-mineral enrichment occurred in the upper swash zone. Sands of the swash zone contain gently southwest-dipping laminae, fine upward, and show local inverse grading. Vos and Hobday proposed a storm-wave microtidal environment prograding within a deltaic depositional regime.

Behr (1986) reports the economic potential of the Bothaville and related deposits to be low, particularly because of problems with recovery of altered ilmenite. Strong cementation of the host rock requires vigorous crushing that destroys friable altered ilmenite.

ECONOMIC PROGNOSIS

Unconsolidated shoreline placer deposits are equally attractive regardless of age, as long as their mineral assemblages and grades are appropriate and overburden is minor. The older deposits may be even more beneficiated by postdepositional weathering or may represent deposition in a paleoclimate more conducive to intensive weathering (Chapter 6). Since the proportion of pre-Quaternary deposits that has been discovered is probably small, they may eventually assume a larger share of world titanium-mineral supply if Quaternary deposits approach exhaustion.

Indurated deposits, on the other hand, must be very high grade to be economically competitive. The requisite grades are documented only in beach deposits, and in these, individual high-grade portions are typically only 1 to 3 m thick, 50 to 150 m wide, and less than 5 km long. Thus they typically contain less than one million tons of heavy minerals. The cementation of the deposits, coupled with the alteration of contained ilmenite, also creates a separation problem; many of the ilmenite grains would not survive crushing.

An intermediate situation in Cretaceous black sandstones of the western United States has been described by Houston and Murphy (1962). Here cementation is in part a surface phenomenon, caused by alteration of detrital iron-titanium oxides at the present land surface. Below this surface, the deposit is more friable, and the cement is partly carbonate and easily removed chemically.

METHODS OF EXPLORATION

Exploration for pre-Quaternary placer deposits of titanium oxide minerals should be aimed at near-surface unconsolidated shoreline placers, as these present the favorable combination of low mining costs and potentially high grades throughout mineable tonnages. Probably a great many of these deposits remain to be discovered.

The conceptual and physical exploration tools that are useful for Quaternary shoreline deposits (Chapter 9) would mostly be useful for unindurated near-surface pre-Quaternary deposits also. The important tool lacking in pre-Quaternary deposits is geomorphic expression, which has been important in exploration for Quaternary deposits. Depositional landforms reveal favorable depositional environments and former fluvial conduits, which are useful in predicting deposit locations. In exploration for pre-Quaternary deposits, this step is blind. For deposits that have been buried, the exploration geologist must use stratigraphic evolution as a substitute for depositional landforms.

Pre-Quaternary beach placers apparently form in prograding shoreline sequences. Thus, exploration should focus on sandy transitions from underlying marine sections to overlying nonmarine sections, in sequences that fill the source-rock, weathering, and conduit requirements of placer formation. In many areas, existing stratigraphic information is sufficient to pinpoint the more favorable shoreline depositional environments in such transitions.

Certain geologic periods are characterized by deep weathering, both regionally and worldwide (Frakes, 1979). Sections representing these periods are everywhere more favorable for formation of economic placer deposits of titanium oxide minerals. Chapter 6 discusses the prediction of favorable depositional regions within favorable time envelopes.

All geophysical methods useful in exploration for Quaternary deposits should be useful for analogous pre-Quaternary ones. Radioactivity surveys deserve special mention because most pre-Quaternary placer deposits, including those described in this chapter, were found by radiometric methods. One reason is historical; many sandstones were examined for uranium in the 1950s. The high monazite contents of the pre-Cenozoic placers enhances their radioactivity.

REFERENCES CITED

Abdullah, M. I., and Atherton, M. P., 1964, The thermometric significance of magnetite in low-grade metamorphic rocks: American Journal of Science, v. 262, p. 904–917.

Achalabhuti, C., Isarangkoon, P., Ratanawong, S., Aeo-Phanthong, V., Kulvanich, S., and Suwanasing, A., 1975, Heavy minerals in southern Thailand: Thailand Department of Mineral Resources Economic Geology Bulletin 9, 106 p.

Adams, J., Zimpfer, G. L., and McLane, C. F., 1978, Basin dynamics, channel processes and placers, and placer formation; A model study: Economic Geology, v. 73, p. 416–426.

Alves, B. P., 1960, Distrito niobio-titanifero de Tapira: Brazil Divisão de Fomento da Produçao Mineral Boletim 108, 48 p.

Anderson, A. T., Jr., and Morin, M., 1969, Two types of massif anorthosites and their implications regarding the thermal history of the crust, in Isachsen, Y. W., ed., Origin of anorthosite and related rocks: New York State Museum and Science Service Memoir 18, p. 57–69.

Arthur, M. A., Dean W. E., and Schlanger, S. O., 1985, Variations in the global carbon cycle during the Cretaceous related to climate, volcanism, and changes in atmospheric CO_2, in Sundquist, E. T., and Broecker, W. S., eds., The carbon cycle and atmospheric CO_2; Natural variations, Archean to present: American Geophysical Union Geophysical Monograph 32, p. 504–546.

Ashwal, L. D., 1982a, Mineralogy of mafic and Fe-Ti oxide-rich differentiates of the Marcy anorthosite massif, Adirondacks, New York: American Mineralogist, v. 67, p. 14–27.

—— , 1982b, Proterozoic anorthosite massifs; A review: Lunar and Planetary Institute Report 82-01, p. 40–45.

Attanasi, E. D., de Young, J. H., Force, E. R., and Grosz, A. E., 1987, Resource assessments, geologic deposit models, and offshore minerals, with an example of heavy-mineral sands, in Teleki, P. G., ed., Marine minerals: London, Reidel, p. 485–513.

Austin, S. R., 1960, Ilmenite, magnetite, and feldspar alteration under reducing conditions: Economic Geology, v. 55, p. 1758–1759.

Badham, J.P.N., and Morton, R. D., 1976, Magnetite-apatite intrusions and calk-alkaline magmatism, Cansell River, N.W.T.: Canadian Journal of Earth Science, v. 13, p. 348–354.

Bailey, S. W., Cameron, E. N., Spedden, H. R., and Weege, R. J., 1956, The alteration of ilmenite in beach sands: Economic Geology, v. 51, p. 263–279.

Balasundaram, S.M.S., 1970, Geological and mineral map of Kerala: Geological Survey of India, scale 1:1,000,000.

Balsley, J. R., Jr., 1943, Vanadium-bearing magnetite-ilmenite deposits near Lake Sanford, Essex County, New York: U.S. Geological Survey Bulletin 940-D, p. 99–123.

Barbosa, O., Braun, O.P.G., Dyer, R. C., and Cunha, C.A.B.R., 1970, Geologia da Região do Triángulo Mineiro: Departamento Nacional Produçao Mineral (Brazil) Boletim 136, 140 p.

Basu, A., and Molinari, E., 1989, Provenance characteristics of detrital opaque Fe-Ti oxide minerals: Journal of Sedimentary Petrology, v. 59, p. 922–934.

Bateman, A. M., 1951, The formation of late magmatic oxide ores: Economic Geology, v. 46, p. 404–426.

Baxter, J., 1972, The geology of the Eneabba area, western Australia, in Report of the Department of Mines, western Australia, for the year 1971: Perth, Western Australia Department of Mines, p. 105–106.

—— , 1977, Heavy mineral sand deposits of western Australia: Geological Survey of Western Australia Mineral Resources Bulletin 10, 148 p.

—— , 1982, Mineral sands mining in Western Australia, in Exploitation of mineral sands: Perth, Western Australia Institute of Technology, p. 61–80.

—— , 1986, Supergene enrichment of ilmenite; Is it related to lateralization? in Australia; A world source of ilmenite, rutile, monazite, and zircon: Perth, Australasian Institute of Mining and Metallurgy, p. 23–27.

Beasley, W. F., 1948, Heavy-mineral beach sands of southern Queensland; Part 1, The nature, distribution and extent, and manner of formation of the deposit: Proceedings of the Royal Society of Queensland, v. 59, p. 109–140.

—— , 1950, Heavy-mineral beach sands of southern Queensland; Part 2, Physical and mineralogical composition, mineral description, and origin of the minerals: Proceedings of the Royal Society of Queensland, v. 61, p. 59–104.

Beck, W. W., Jr., 1973, Correlation of Pleistocene barrier islands in the lower Coastal Plain of South Carolina as inferred by heavy minerals: South Carolina Division of Geology Geologic Notes, v. 17, no. 3, p. 55–67.

Behr, S. H., 1965, Heavy-mineral beach deposits in the Karoo System: South African Geological Survey Memoir 56, 116 p.

—— , 1986, Heavy mineral deposits in the Karoo Sequence, South Africa, in Anhaesser, C. R., ed., Mineral deposits of southern Africa: Johannesburg, Geological Society of South Africa, p. 2105–2118.

Bergeron, M., 1972, Quebec Iron and Titanium Corporation ore deposit at Lac Tio, Quebec: International Geological Congress, 1972, Excursion guidebook B 09, 8 p.

—— , 1986, Minéralogie et géochimie de la suite anorthositique de la région du Lac Allard, Québec: Évolution des membres mafiques et origine des gites massifs d'ilménite [Ph.D. thesis]: University of Montreal, 480 p.

Bergstøl, S., 1972, The jacupirangite at Kodal, Vestfold, Norway: Mineralium Deposita, v. 7, p. 233–246.

Berner, R. A., and Landis, G. P., 1988, Gas bubbles in fossil amber as possible indicators of the major gas composition of ancient air: Science, v. 239, p. 1406–1409.

Berner, R. A., Lasaga, A. C., and Garrels, R. M., 1983; The carbonate-silicate geochemical cycle and its effect on atmospheric carbon dioxide over the last 100 million years: American Journal of Science, v. 283, p. 641–683.

Beurlen, H., and Cassedanne, J. P., 1981, The Brazilian mineral resources: Earth Science Reviews, v. 17, p. 177–206.

Bigarella, J. J., Becker, R. D., and Duarte, G. M., 1969, Coastal dune structures from Parana (Brazil): Marine Geology, v. 7, p. 5–55.

Binns, R. A., 1967, Barroisite-bearing eclogite from Naustdal, Sogn og Fjordane, Norway: Journal of Petrology, v. 8, p. 349–371.

Black, P. M., 1977, Regional high-pressure metamorphism in New Caledonia: Phase equilibria in the Ouegoa district: Tectonophysics, v. 43, p. 89–107.

Blake, M. C., Jr., and Morgan, B. A., 1976, Rutile and sphene in blueschist and related high-pressure facies rocks: U.S. Geological Survey Professional Paper 959C, p. C1–C6.

Bolsover, L. R., and Lindsley, D. H., 1983, Sybille oxide deposit; Massive Fe-Ti oxides intrusive into the Laramie Anorthosite Complex, Wyoming [abs.]: EOS Transactions of the American Geophysical Union, v. 64, no. 18, p. 328.

Bradley, J. S., 1957, Differentiation of marine and subaerial sedimentary environments by volume percentage of heavy minerals, Mustang Island, Texas: Journal of Sedimentary Petrology, v. 27, p. 116–125.

Bramall, A., 1928, Dartmoor detritals; A study of provenance: Proceedings of the Geological Association, v. 39, p. 27–48.

Bray, R. E., and Wilson, J. C., 1975, Guidebook to the Bingham mining district: Society of Economic Geologists, 156 p.

Brøgger, W. C., 1934–1935, On the several Archean rocks from the south coast of Norway; 2, The south Norwegian hyperites and their metamorphism: Oslo, Norsk Videnskaps-Academi Oslo Skrifter, Section I. Matem.-Natur Klasse, 421 p.

Brown, C. E., 1983, Mineralization, mining, and mineral resources in the Beaver Creek area of the Grenville lowlands in St. Lawrence County, New York: U.S. Geological Survey Professional Paper 1279, 21 p.

Buddington, A. F., Fahey, J., and Vlisidis, A., 1955, Thermometric and petrogenetic significance of titaniferous magnetite: American Journal of Science, v. 253, p. 497–532.

Cameron, E. N., 1979, Titanium-bearing oxide mineral of the Critical Zone of the eastern Bushveld Complex: American Mineralogist, v. 67, p. 140–150.

Canadian Mining Journal, 1956, Beach sand mining from the sands of Travancore: Canadian Mining Journal, v. 77, p. 67–76.

Cannon, H. B., 1950, Economic minerals in the beach sands of the southeastern United States, in Snyder, F. G., ed., Proceedings of the Symposium on

Mineral Resources of the southeastern United States: Knoxville, University of Tennessee Press, p. 202–210.

Carmichael, I.S.E., Nicholls, J., and Smith, A. L., 1970, Silica activity in igneous rocks: American Mineralogist, v. 55, p. 246–263.

Carpenter, R. H., 1982, Mineralization and alteration in the Lincolnton, Ga., McCormick, S.C., area, *in* Allard, G. O., and Carpenter, R. H., eds., Exploration for metallic resources in the Southeast: Athens, University of Georgia Department of Geology, p. 120–142.

Carpenter, R. H., and Allard, G. O., 1982, Aluminosilicate assemblages; An exploration tool for metavolcanic terrains of the southeast, *in* Allard, G. O., and Carpenter, R. H., eds., Exploration for metallic resources in the Southeast: Athens, University of Georgia Department of Geology, p. 19–22.

Carroll, D., 1939, Beach sands from Bunbury, Western Australia: Journal of Sedimentary Petrology, v. 9, p. 95–107.

Carroll, D., Neuman, R. B., and Jaffe, H. W., 1957, Heavy minerals in arenaceous beds in parts of the Ocoee Series, Great Smoky Mountains, Tennessee: American Journal of Science, v. 255, p. 175–193.

Carson, D.J.T., and Jambor, J. L., 1974, Mineralogy, zonal relationships, and economic significance of hydrothermal alteration at porphyry copper deposits, Babine Lake area, British Columbia: Canadian Mining and Metallurgical Bulletin, v. 67, p. 110–133.

Carter, B. A., 1982a, Geology and structural setting of the San Gabriel anorthosite–syenite body and adjacent rocks of the western San Gabriel Mountains, Los Angeles County, California, *in* Geologic excursions in the Transverse Ranges, southern California; Geological Society of America, Cordilleran Section, Guidebook Anaheim meeting, Trips 5, 6, 11: p. 1–56.

—— , 1982b, Mineral potential of the San Gabriel anorthosite–syenite body, San Gabriel Mountains, California, *in* Fife, D. L., and Minch, J. A., eds., Geology and mineral wealth of the California ranges: Santa Ana, California, South Coast Geological Society, p. 208–212.

Carter, C. H., 1978, A regressive barrier and barrier-protected deposit; Depositional environments and geographic setting of the late Tertiary Cohansey Sand: Journal of Sedimentary Petrology, v. 48, p. 933–950.

Carvalho, W. T., 1974, Aspectos geológicos e petrográficos do complexo ultramáfico-alcalino de Catalão I, Goias: Pôrto Alegre, Sociedad Braziliera de Geologia, Annais do 28th Congresso, v. 4–5, p. 107–123.

Cazeau, C. J., 1974, Heavy minerals of Quaternary sands in South Carolina, *in* Oaks, R. Q., Jr., and Du Bar, J. R., eds., Post-Miocene stratigraphy, central and southern Atlantic Coastal Plain: Logan, Utah State University Press, p. 174–178.

Centro de Investigacion Mineria y Metalurgia, 1986, Recuperacion de rutilo desde relaves de cobre porfidico: Boletin, v. 1, 4 p.

Chappell, B. W., and White, A.J.R., 1974, Two contrasting granite types: Pacific Geology, v. 8, p. 173–174.

Chenowith, W. L., 1957, Radioactive titaniferous heavy-mineral deposits in the San Juan basin, New Mexico and Colorado, *in* Southwestern San Juan Mountains Guidebook: New Mexico Geological Society 8th Field Conference Guidebook, p. 212–217.

Chesnokov, B. V., 1960, Rutile bearing eclogites from the Shubino village deposit in the southern Urals: International Geology Review, v. 2, p. 936–945.

Chidester, A. H., 1962, Petrology and geochemistry of selected talc-bearing ultramafic rocks and adjacent country rocks in north-central Vermont: U.S. Geological Survey Professional Paper 345, 207 p.

Clerici, C., Mancini, A., Mancini, R., Morandini, A., Occella, E., and Protlo, C., 1981, Recovery of rutile from an eclogite rock, *in* Laskowki, J., ed., Mineral processing; Developments in mineral processing: New York, Elsevier, v. 2, pt. B, p. 1803–1827.

Clifton, H. E., 1969, Beach lamination; Nature and origin: Marine Geology, v. 7, p. 553–559.

Clifton, H. E., Hunter, R. E., and Phillips, R. S., 1971, Depositional structures and processes in the non-barred high-energy nearshore: Journal of Sedimentary Petrology, v. 41, p. 651–670.

Coleman, R. G., Lee, D. E., Beatty, L. B., and Brannock, W. W., 1965, Eclogites

and eclogites—their differences and similarities: Geological Society of America Bulletin, v. 76, p. 483–508.

Collins, L. B, and Baxter, J. L., 1984, Heavy mineral-bearing strandline deposits associated with high-energy beach environments, southern Perth Basin, Western Australia: Australian Journal of Earth Sciences, v. 31, p. 287–292.

Collins, L. B., and Hamilton, N.T.M., 1986, Stratigraphic evolution and heavy-mineral accumulation in the Minninup shoreline, southwest Australia, *in* Australia; A world source of ilmenite, rutile, monazite, and zircon: Perth, Australasian Institute of Mining and Metallurgy, p. 17–22.

Collins, L. B., Hochwimmer, B., and Baxter, J. L., 1986, Depositional facies and mineral deposits of the Yoganup shoreline, southern Perth basin, *in* Australia; A world source of ilmenite, rutile, monazite, and zircon: Perth, Australasian Institute of Mining and Metallurgy, p. 9–16.

Colquhoun, D. J., 1965, Terrace sediment complexes in central South Carolina: Atlantic Coastal Plain Geological Association Field conference 1965, 62 p.

Colwell, J. B., 1982a, Heavy mineral data from east Australian coastal sediments, Newcastle to Fraser Island: Geologische Jahrbuch, v. D 56, p. 49–53.

—— , 1982b, Sedimentology of surface sediments of the New South Wales shelf: Geologische Jahrbuch, v. D 56, p. 111–124.

Connah, T. H., 1961, Beach sand heavy mineral deposits of Queensland: Geological Survey of Queensland Publication 302, 31 p.

Cortesogno, L., Ernst, W. G., Galli, M., Messiga, B., Pedemonte, G. M., and Piccardo, G. B., 1977, Chemical petrology of eclogitic lenses in serpentinite, Gruppo di Voltri, Ligurian Alps: Journal of Geology, v. 85, p. 255–277.

Crawford, E. A., Herbert, C., Taylor, G., Helby, R., Morgan, R., and Ferguson, J., 1980, Diatremes of the Sydney Basin: Geological Survey of New South Wales Bulletin 26, p. 294–323.

Creitz, E. E., and McVay, T. N., 1948, A study of opaque minerals in Trail Ridge, Florida, dune sands: American Institute of Mining and Metallurgical Engineers Technical Publication 2426, 7 p.

Cressman, E. R., 1962, Nondetrital siliceous sediments: U.S. Geological Survey Professional Paper 440-T, 22 p.

Crowder, D. F., and Sheridan, M. F., 1972, Geologic map of the White Mountain Peak Quadrangle, Mono County, California: U.S. Geological Survey Geologic Quadrangle Map GQ-1012, scale 1:62,500.

Cruz, F. T., Machado, P. T., Townsend, R., and Machado, J. E., 1977, Aspects of the superficial phosphorus and titanium deposits of the Tapira Carbonatite Complex, Minas Gerais, Brazil: 25th International Geological Congress, Sydney, Abstracts, v. 1, p. 209–210.

Czamanske, G. K., Force, E. R., and Moore, W. J., 1981, Some geologic and potential resources aspects of rutile in porphyry copper deposits: Economic Geology, v. 76, p. 2240–2245.

Danilchik, W., and Haley, B. R., 1964, Geology of the Paleozoic area in the Malvern Quadrangle, Garland and Hot Springs Counties, Arkansas: U.S. Geological Survey Map I-405, scale 1:48,000.

Darby, D. A., 1984, Trace elements in ilmenite; A way to discriminate provenance or age in coastal sands: Geological Society of America Bulletin, v. 95, p. 1208–1218.

Darby, D. A., and Tsang, Y.-W., 1987, Variation in ilmenite element composition within and among drainage basins; Implications for provenance: Journal of Sedimentary Petrology, v. 57, p. 831–838.

Davidson, D., 1982, Exploration methods for mineral sands, *in* Exploitation of mineral sands: Perth, Western Australia Institute of Technology, p. 107–122.

Davies, O., 1970, Pleistocene beaches of Natal: Annals of the Natal Museum, v. 20, p. 403–442.

Davis, E. G., and Sullivan, G. V., 1971, Recovery of heavy minerals from sand and gravel operations in the southeastern United States: U.S. Bureau of Mines Report of Investigation 7517, 25 p.

Davis, E. G., Sullivan, G. V., and Lamont, W. E., 1988, Potential for recovery of rutile and other by-products from western copper tailings: U.S. Bureau of Mines Report of Investigation 9158, 19 p.

Dennen, W. H., and Anderson, P. J., 1962, Chemical changes in incipient rock weathering: Geological Society of America Bulletin, v. 73, p. 375–384.

Desborough, G. A., and Mihalik, P., 1980, Accessory minerals in the igneous host of molybdenum ore, Henderson Mine, Colorado: U.S. Geological Survey Open-File Report 80–661, 16 p.

Desborough, G. A., and Sharp, W. N., 1978, Tantalum, uranium, and scandium in heavy accessory oxides, Climax molybdenum mine, Climax, Colorado: Economic Geology, v. 73, p. 1749–1751.

Diemer, R. A., 1941, Titaniferous magnetite deposits of the Laramie Range, Wyoming: Geological Survey of Wyoming Bulletin 31, 23 p.

Dimanche, F., 1972, Évolution minéralogique de quelques sables titaniferes d'Afrique du Sud: Annales Société Géologique de Belgique, t. 95, p. 183–190.

Dimanche, F., and Bartholome, P., 1976, The alteration of ilmenite in sediments: Minerals Science Engineering, v. 8, p. 187–200.

Dow, V. T., 1961, Magnetite and ilmenite resources, Iron Mountain area, Albany County, Wyoming: U.S. Bureau of Mines Information Circular 8037, 133 p.

Dow, V. T., and Batty, J. V., 1961, Reconnaissance of titaniferous sandstone deposits of Utah, Wyoming, New Mexico, and Colorado: U.S. Bureau of Mines Report of Investigation 5860, 52 p.

Dryden, L., and Dryden, C., 1946, Comparative rates of weathering of some common heavy minerals: Journal of Sedimentary Petrology, v. 16, p. 91–96.

Duane, D. B., Field, M. E., Meisburger, E. P., Swift, D.J.P., and Williams, S. J., 1972, Linear shoals on the Atlantic Inner Continental Shelf, Florida, to Long Island, *in* Swift, D.J.P., Duane, D. B., and Pilkey, O. H., eds., Shelf sediment transport: Process and pattern: Stroudsburg, Pennsylvania, Dowden, Hutchinson and Ross, p. 499–575.

Duchesne, J. C., 1972, Iron-titanium oxide minerals in the Bjerkrem–Sogndal massif, southwestern Norway: Journal of Petrology, v. 13, p. 57–81.

Duchesne, J. C., Denoiseux, B., and Hertogen, J., 1987, The norite-mangerite relations in the Bjerkreim–Sokndal layered lopolith (southwest Norway): Lithos, v. 20, p. 1–17.

Dymek, R. F., 1983, Titanium, aluminum, and interlayer cation substitutions in biotite from high-grade gneisses, West Greenland: American Mineralogist, v. 68, p. 880–899.

Eberle, M. C., and Atkinson, W. W., 1983, Results of mapping at Iron Mountain, Laramie Anorthosite Complex, Wyoming: Geological Society of America Abstracts with Programs, v. 15, p. 565.

Elger, G. W., Wright, J. B., Tress, J. E., Bell, H. E., and Jordan, R. R., 1986, Producing chlorination-grade feedstock from domestic ilmenite-laboratory and pilot-plant studies: U.S. Bureau of Mines Report of Investigation 9002, 24 p.

Elliot, C. L., and Guilbert, J. M., 1975, Induced polarization response attributed to magnetite and certain Fe-Ti oxide minerals: Geophysics, v. 40, p. 147.

Elsdon, R., 1975, Iron-titanium oxide minerals in igneous and metamorphic rocks: Minerals Science Engineering, v. 1, p. 48–70.

Emslie, R. F., 1978, Anorthosite massifs, rapakivi granites, and late Proterozoic rifting of North America: Precambrian Research, v. 7, p. 61–98.

Epler, N. E., Bolsover, L. R., and Lindsley, D. H., 1986, Nature and origin of the Sybille Fe-Ti oxide deposit, Laramie Anorthosite complex, SE Wyoming: Geological Society of America Abstracts with Programs, v. 18, p. 595.

Erickson, R. L., and Blade, L. V., 1963, Geochemistry and petrology of the alkalic igneous complex at Magnet Cove, Arkansas: U.S. Geological Survey Professional Paper 425, 95 p.

Ernst, W. G., 1976, Mineral chemistry of eclogites and related rocks from the Voltri Group, western Liguria, Italy: Schweizerishe Mineralogishe and Petrographishe Mitteilungen, v. 56, p. 293–343.

Ernst, W. G., Seki, Y., Onuki, H., and Gilbert, M. C., 1970, Comparative study of low-grade metamoprhism in the California coast ranges and the outer metamorphic belt of Japan: Geological Society of America Memoir 124, 276 p.

Eskola, P., 1921, On the eclogites of Norway: Skrifter, Videnskapsselskapet i Christiania, Matematisk–Naturvidenskabelig Klasse I, v. 8, 118 p.

Espenshade, G. H., and Potter, D. B., 1960, Kyanite, sillimanite, and andalusite deposits of the southeastern states: U.S. Geological Survey Professional Paper 336, 121 p.

Fairchild, H. S., 1912, The glacial waters in the Black and Mohawk valleys: New York State Museum Bulletin 160, 47 p.

Falkum, T., 1982, Geologisk kart over Norge, berggrunnskart "Mandal": Norges geologiske undersokelse (Norway), scale 1:250,000.

Fantel, R. J., Buckingham, D. A., and Sullivan, D. E., 1986, Titanium minerals availability; Market economy countries: U.S. Bureau of Mines Information Circular 9061, 48 p.

Ferguson, C. C., and Garman, R. K., 1970, Mineral resources summary of the Buchanan Quadrangle, Tennessee: Tennessee Division of Geology and Mineral Resources Summary MRS 19-NW (and map), 16 p.

Field, M. E., and Roy, P. S., 1985, Offshore transport and sand-body formation; Evidence from a steep, high-energy shoreface, southeastern Australia: Journal of Sedimentary Petrology, v. 54, p. 1292–1302.

Filho, J.G.C.S., 1974, Prospecção de urânio nas chaminés alcalinas de Serra Negra e Salitre: Comissão Nacional de Energia Nuclear (Brazil) Boletim 9, 41 p.

Fischer, A. G., 1981, Climatic oscillations in the biosphere, *in* Nitecki, M., ed., Biotic crises in ecological and evolutionary time: New York, Academic, p. 101–131.

Fish, G. E., 1962, Titanium resources of Nelson and Amherst Counties, Virginia; Part 1, Saprolite ores: U.S. Bureau of Mines Report of Investigation 6094, 44 p.

Fleischer, M., Murata, K. J., Fletcher, J. D., and Narten, P. F., 1952, Geochemical association of niobium (columbium) and titanium, and its geological and economic significance: U.S. Geological Survey Circular 225, 13 p.

Flemming, B. W., 1981, Factors controlling shelf sediment dispersal along the southeast African continental margin: Marine Geology, v. 42, p. 259–277.

Flinter, B. H., 1959, The alteration of Malayan ilmenite grains and the question of "arizonite": Economic Geology, v. 54, p. 720–729.

Flohr, M.J.K., and Ross, M., 1989, Alkaline igneous rocks of Magnet Cove, Arkansas; Metasomatized ijolite xenoliths from Diamond Jo Quarry: American Mineralogist, v. 74, p. 113–131.

Fockema, P. D., 1986, The heavy mineral deposits north of Richards Bay, *in* Anhaensser, C. R., ed., Mineral deposits of southern Africa: Johannesburg, Geological Society of South Africa, p. 2301–2307.

Force, E. R., 1976a, Titanium contents and titanium partitioning in rocks: U.S. Geological Survey Professional Paper 959-A, 10 p.

——, 1976b, Metamorphic source rocks of titanium placer deposits; A geochemical cycle: U.S. Geological Survey Professional Paper 959 B, 16 p.

——, 1976c, Titanium minerals in deposits of other minerals: U.S. Geological Survey Professional Paper 959-F, 5 p.

——, 1980a, Is the United States of America *geologically* dependent on imported rutile? Proceedings of the Industrial Minerals, 4th Industrial Minerals International Congress, p. 43–47.

——, 1980b, The provenance of rutile: Journal of Sedimentary Petrology, v. 50, p. 485–488 (reprinted 1985, *in* Leupke, G., ed., Economic analysis of heavy minerals in sediments: New York, Van Nostrand Reinhold, p. 230–233).

Force, E. R., and Carter, B. A., 1986, Liquid immiscibility proposed for nelsonitic components of the anorthosite-syenite-gabbro complex, San Gabriel Mountains, California: Geological Society of America Abstracts with Programs, v. 18, p. 604.

Force, E. R., and Garnar, T. E., Jr., 1985, High angle aeolian crossbedding at Trail Ridge, Florida: Industrial Minerals, August, p. 55–59.

Force, E. R., and Geraci, P. J., 1975, Heavy minerals in Pleistocene(?) shoreline sand bodies of southeastern Virginia: U.S. Geological Survey Map MF-718, scale 1:250,000.

Force, E. R., and Lynd, L. E., 1984, Titanium-mineral resources of the United States; Definitions and documentation: U.S. Geological Survey Bulletin 1558-B, 11 p.

Force, E. R., and Rich, F. J., 1989, Geologic evolution of Trail Ridge eolian heavy mineral sand and underlying peat, northern Florida: U.S. Geological Survey Professional Paper 1499, 16 p.

Force, E. R., and Stone, B. D., 1990, Heavy mineral dispersal and deposition in sandy deltas of glacial Lake Quinebaug, Connecticut: U.S. Geological Survey Bulletin 1874, 21 p.

Force, E. R., Lipin, B. R., and Smith, R. E., 1976, Heavy mineral resources in Pleistocene sand of the Port Leyden Quadrangle, southwestern Adirondack Mountains, New York: U.S. Geological Survey Map MF-728-B, scale 1:24,000.

Force, E. R., Grosz, A. E., Loferski, P. J., and Maybin, A. H., 1982, Aeroradioactivity maps in heavy-mineral exploration; Charleston, South Carolina: U.S. Geological Survey Professional Paper 1218, 19 p.

Force, E. R., Sukirno, D., and van Leeuwen, T., 1984, Exploration for porphyry metal deposits based on rutile distribution; A test in Sumatera: U.S. Geological Survey Bulletin 1558-A, 9 p. (also Indonesia Directorate of Mineral Resources Special Publication, v. 1, p. 1–13).

Frakes, L. A., 1979, Climates through geologic time: Amsterdam, Elsevier, 310 p.

Friis, H., 1974, Weathered heavy-mineral associations from the young Tertiary deposits of Jutland, Denmark: Sedimentary Geology, v. 12, p. 199–213.

Frost, M. T., Grey, I. E., Harrowfield, I. R., and Mason, K., 1983, The dependence of alumina and silica contents on the extent of alteration of weathered ilmenites from Western Australia: Mineralogical Magazine, v. 47, p. 201–208.

Fryklund, V. C., and Holbrook, D. F., 1950, Titanium ore deposits of the Magnet Cove area, Hot Springs County, Arkansas: Arkansas Division of Geology Bulletin 16, 173 p.

Galloway, M. C., 1972, Statistical analyses of regional heavy mineral variation, Hawkesbury sandstone and Narrabeen Group (Triassic), Sydney Basin: Geological Society of Australia Journal, v. 19, p. 65–76.

Gardner, D. E., 1955, Beach-sand heavy-mineral deposits of eastern Australia: Bureau of Mineral Resources Bulletin 28, 103 p.

Garnar, T. E., 1972, Economic geology of Florida heavy-mineral deposits: Florida Bureau of Geology Special Publication 17, p. 17–21.

———, 1978, Heavy mineral mining in Florida [abs.]: American Institute of Mining, Metallurgical, and Petroleum Engineers Technical Program, p. 19.

———, 1980, Heavy minerals industry of North America, *in* Coope, B. M., ed., Proceedings of the 4th Industrial Minerals International Congress: London, Industrial Minerals, p. 29–42.

Giejer, P., 1964, Genetic relationships of the paragenesis Al_2SiO_5-lazulite-rutile: Arkiv Mineralogi och Geologi, v. 3, p. 423–466.

Geisel-Sobrinho, E., 1974, Prospecção de urânio na chaminé alcalina de Tapira, Minas Gerais: Comissão Nacional de Energia Nuclear (Brazil) Boletim 10, 16 p.

Ghent, E. D., and Stout, M. Z., 1984, TiO_2 activity in metamorphosed pelitic and basic rocks; Principles and applications to metamorphism in southeastern British Columbia: Contributions to Mineralogy and Petrology, v. 86, p. 248–255.

Gierth, E., and Krause, H., 1973, Die Ilmenitlagerstatte Tellnes (süd-Norwegen): Norsk Geologiske Tidsskrift, v. 53, p. 359–402.

———, 1974, Contribution to the mineralogy of Norway, no. 57: Baddelyit von Tellnes: Norsk Geologiske Tidsskrift, v. 54, p. 193–197.

Gillson, J. L., 1959, Sand deposits of titanium minerals: Mining Engineering, v. 11, p. 421–429.

Goldberg, S. A., 1984, Geochemical relationships between anorthosite and associated iron-rich rocks, Laramie Range, Wyoming: Contributions to Mineralogy and Petrology, v. 87, p. 376–387.

Goldich, S. S., 1938, A study in rock weathering: Journal of Geology, v. 46, p. 17–58.

Goldsmith, R., and Force, E. R., 1978, Distribution of rutile in metamorphic rocks and implications for placer deposits: Mineralium Deposita, v. 13, p. 329–343.

Gomes, J. M., Martinez, G. M., and Wong, M. M., 1979, Recovering by-product heavy minerals from sand and gravel, placer gold, and industrial mineral operations: U.S. Bureau of Mines Report of Investigation 8366, 15 p.

———, 1980, Recovery of by-product heavy minerals from sand and gravel opera-

tions in central and southern California: U.S. Bureau of Mines Report of Investigation 8471, 20 p.

Gottfried, D., Greenland, P. L., and Campbell, E. Y., 1968, Variation of Nb-Ta, Zr-Hf, Th-U, and K-Cs in two diabase-granophyre suites: Geochimica et Cosmochimica Acta, v. 32, p. 925–947.

Green, J. C., 1956, Geology of the Storkollen–Blankenburg area, Kragerø, Norway: Norsk Geologisk Tidsskrift, v. 36, p. 89–140.

Grey, I. E., and Reid, A. F., 1975, The structure of pseudorutile and its role in the natural alteration of ilmenite: American Mineralogist, v. 60, p. 898–906.

Griffin, W. L., 1987, On the eclogites of Norway—65 years later: Mineralogical Magazine, v. 51, p. 333–343.

Griffin, W. L., and Mork, M.B.E., 1981, Eclogites and basal gneisses in western Norway, Excursion B1: Oslo, Geologisk Museum, 88 p.

Griffin, W. L., and 8 others, 1985, High-pressure metamorphism in the Scandinavian Caledonides, *in* The Caledonide Orogen; Scandinavia and related areas: New York, Wiley, p. 783–801.

Grigg, N. S., and Rathbun, R. E., 1969, Hydraulic equivalence of minerals with a consideration of the reentrainment process: U.S. Geological Survey Professional Paper 650-B, p. B77–B80.

Grobler, N. J., and Whitfield, G. G., 1970, The olivine-apatite magnetites and selected rocks in the Villa Nora occurrence of the Bushveld Igneous Complex: Geological Society of South Africa Special Publication 1, p. 208–227.

Gross, E. B., and Parwell, A., 1969, Rutile mineralization at the White Mountain andalusite deposits, California: Arkiv Mineralogi och Geologi, v. 4, paper 29, p. 493–497.

Gross, S. O., 1968, Titaniferous ores of the Sanford Lake district, New York, *in* Ridge, J. D., ed., Ore deposits of the United States, 1933–1967, v. 1: New York, American Institute of Mining, Metallurgical, and Petroleum Engineers, p. 140–153.

Grosz, A. E., 1983, Application of total-count aeroradiometric maps to the exploration for heavy mineral deposits in the coastal plain of Virginia: U.S. Geological Survey Professional Paper 1263, 20 p.

———, 1987, Nature and distribution of potential heavy-mineral resources offshore of the Atlantic coast of the United States: Marine Mining, v. 6, p. 339–357.

Grosz, A. E., Hathaway, J. C., and Escowitz, E. C., 1986, Placer deposits of heavy minerals in Atlantic continental shelf sediments: Offshore technology conference, 18th, Houston, Preprint OTC 5196, p. 387–392.

Grout, F. F., 1949–1950, The titaniferous magnetites of Minnesota: St. Paul, Office of the Commissioner of the Iron Range Resources and Rehabilitation, 117 p.

Groves, A. W., 1931, The unroofing of the Dartmoor Granite and the distribution of its detritus in the sediments of southern England: Quarterly Journal of the Geological Society, v. 87, p. 62–96.

Guidotti, C. V., Cheney, J. T., and Guggenheim, S., 1977, Distribution of titanium between coexisting muscovite and biotite in pelitic schists from northwestern Maine: American Mineralogist, v. 62, p. 438–448.

Guild, P. W., 1971, Metallogeny; A key to exploration: Mining Engineering, v. 23, p. 69–72.

Gustavson, T. C., Ashley, G. M., and Boothroyd, J. C., 1975, Depositional sequences in glaciolacustrine deltas, *in* Jopling, A. V., and McDonald, B. C., eds., Glaciofluvial and glaciolacustrine sedimentation: Society of Economic Paleontologists and Mineralogists Special Publication 23, p. 264–280.

Haggerty, S. E., 1976a, Opaque mineral oxides in terrestrial igneous rocks, *in* Rumble, D., III, ed., Oxide minerals: Mineralogical Society of America Short Course Notes, v. 3, p. Hg 101–300.

———, 1976b, Oxidation of opaque mineral oxides in basalts, *in* Rumble, D., III, ed., Oxide minerals: Mineralogical Society of America Short Course Notes, v. 3, p. Hg 1–100.

Hagner, A. F., 1968, The titaniferous magnetite deposit at Iron Mountain, Wyoming, *in* Ridge, J. D., ed., Ore deposits of the United States, 1933–1967, v. 1: New York, American Institute of Mining, Metallurgical, and Petroleum Engineers, p. 666–680.

Hahn, A. D., and Fine, A. D., 1960, Examination of ilmenite-bearing sands in Otter Creek valley, Kiowa and Tillman Counties, Oklahoma: U.S. Bureau of Mines Report of Investigation 5577, 77 p.

Hails, J. R., 1969, The nature and occurrence of heavy minerals in three coastal areas of New South Wales: Journal of the Royal Society of New South Wales, v. 102, p. 21–39.

Hails, J. R., and Hoyt, J. H., 1972, The nature and occurrence of heavy minerals in Pleistocene and Holocene sediments of the lower Georgia coastal plain: Journal of Sedimentary Petrology, v. 42, p. 646–666.

Hammarbeck, E.C.I., 1976, Titanium, *in* Coetzee, C. B., ed., Mineral resources of the Republic of South Africa: South African Geological Survey Handbook 7, p. 221–226.

Hammond, P., 1952, Allard Lake ilmenite deposits: Economic Geology, v. 47, p. 634–649.

Hand, B. M., 1967, Differentiation of beach and dune sands, using settling velocities of light and heavy minerals: Journal of Sedimentary Petrology, v. 27, p. 514–520.

Harben, P., 1984, Titanium minerals in Brazil; Progress and potential: Industrial Minerals, January, p. 45–49.

Hargraves, R. B., 1962, Petrology of the Allard Lake anorthosite suite, *in* Engel, A.E.J., and others, eds., Petrologic studies: Geological Society of America Buddington Volume, p. 163–189.

Hartley, M. E., 1971, Ultramafic and related rocks in the vicinity of Lake Chatuge: Georgia Geological Survey Bulletin 85, 61 p.

Hartman, J. A., 1959, The titanium mineralogy of certain bauxites and their parent materials: Economic Geology, v. 54, p. 1380–1405.

Hedlund, D. C., and Olson, J. C., 1975, Geologic map of the Powderhorn Quadrangle, Gunnison and Saguache Counties, Colorado: U.S. Geological Survey Map GQ-1178, scale 1:24,000.

Heinrich, E. W., 1966, The geology of carbonatites: Chicago, Rand McNally, 555 p.

Hershey, R. E., 1966, Mineral resources summary of the Seventeen Creek Quadrangle, Tennessee: Tennessee Division of Geology Mineral Resources Summary 21-NW (and map), 17 p.

—— , 1968, Mineral resources summary of the Chesterfield Quadrangle, Tennessee: Tennessee Division of Geology Mineral Resources Summary 11-NE (and map), 24 p.

Herz, N., 1969, The Roseland alkalic anorthosite massif, Virginia, *in* Isachsen, Y. W., ed., Origin of anorthosite and related rocks: New York State Museum and Science Service Memoir 18, p. 357–367.

—— , 1976, Titanium deposits in alkalic igneous rocks: U.S. Geological Survey Professional Paper 959-E, 6 p.

Herz, N., and Force, E. R., 1987, Geology and mineral deposits of the Roseland district of central Virginia: U.S. Geological Survey Professional Paper 1371, 56 p.

Herz, N., and Valentine, E. B., 1970, Rutile in the Harford County, Maryland, serpentinite belt: U.S. Geological Survey Professional Paper 700-C, p. 43–48.

Higgs, D. V., 1954, Anorthosite and related rocks of the western San Gabriel Mountains, southern California: University of California Publications in Geological Sciences, v. 30, p. 171–222.

Hocking, R. M., Warren, J. K., and Baxter, J. L., 1982, Shoreline geomorphology, *in* Exploitation of mineral sands: Perth, Western Australia Institute of Technology, p. 81–92.

Hocq, M., 1982, Region du lac Allard: (Quebec) Ministere de Energie et des Ressources Rapport DPV-894, 99 p.

Holbrook, D. F., 1948, Molybdenum in Magnet Cove, Arkansas: Arkansas Division of Geology Bulletin 12, 16 p.

Hosterman, J. W., Scheid, V. E., Allen, V. T., and Sohn, I. G., 1960, Investigations of some clay deposits in Washington and Idaho: U.S. Geological Survey Bulletin 1091, 147 p.

Houston, R. S., and Murphy, J. F., 1962, Titaniferous black sandstone deposits of Wyoming: Geological Society of Wyoming Bulletin 49, 120 p.

—— , 1977, Depositional environment of Upper Cretaceous black sandstones of the western interior: U.S. Geological Survey Professional Paper 994 A, 29 p.

Howard University Geology Field Camp, 1974, Geologic map of the Port Leyden Quadrangle: New York State Geological Survey Open-File Report, scale 1:24,000.

Hubert, J. F., 1962, A zircon-tourmaline-rutile maturity index and the interdependence of the composition of heavy mineral assemblages with the gross composition and texture of sandstones: Journal of Sedimentary Petrology, v. 32, p. 440–450.

Hudson, J. P., 1986, Evidence for littoral drift at Holocene low sea levels on the south Sydney inner continental shelf, southeastern Australia: 12th International Sedimentological Congress, Canberra, Abstracts, p. 146.

Hunt, J. A., and Kerrick, D. M., 1977, The stability of sphene; Experimental redetermination and geologic implications: Geochimica et Cosmochimica Acta, v. 41, p. 279–288.

Hunter, R. E., 1968, Heavy minerals of the Cretaceous and Tertiary sands of extreme southern Illinois: Illinois Geological Survey Circular 428, 22 p.

Hunter, R. E., and Richmond, B. M., 1988, Daily cycles in coastal dunes: Sedimentary Geology, v. 55, p. 43–67.

Industrial Minerals, 1972, By-product minerals from tin mining: June, p. 35.

—— , 1978, Richards Bay '78, Tapira '88: March, p. 7.

—— , 1986, New heavy minerals supplier: May, p. 14.

Ishihara, S., 1977, The magnetite-series and ilmenite-series granitic rocks: Mining Geology, v. 27, p. 293–305.

Itaya, T., and Banno, S., 1980, Paragenesis of titanium-bearing accessories in pelitic schists of the Sanbagawa metamorphic belt, central Shikoku, Japan: Contributions to Mineralogy and Petrology, v. 73, p. 267–276.

Itaya, T., and Otsuki, M., 1978, Stability and paragenesis of Fe-Ti oxide minerals and sphene in the basis schists of the Sanbagawa metamorphic belt in central Shikoku, Japan: Japanese Association of Mineralogists, Petrologists, and Economic Geologists Journal, v. 73, p. 359–379.

Iwasaki, I., Smith, K. A., and Maliesi, A. S., 1982, By-product recovery from copper-nickel bearing Duluth Gabbro: Minneapolis, University of Minnesota Mineral Resources Research Center, 283 p.

Jackson, M. L., and Sherman, G. D., 1953, Chemical weathering of minerals in soils, *in* Norman, A. G., ed., Advances in agronomy, v. 5: New York, Academic Press, p. 217–319.

Jacob, K., 1956, Ilmenite and garnet sands of the Chowghat, Tinnevelly, Ramnad, and Tanjore coasts: India Geological Survey Records, v. 82, p. 567–602.

Jones, H. A., and Davies, P. J., 1979, Preliminary studies of offshore placer deposits, eastern Australia: Marine Geology, v. 30, p. 243–268.

Karkhanavala, M. D., and Momin, A. C., 1959, The alteration of ilmenite: Economic Geology, v. 54, p. 1095–1102.

Karkhanavala, M. D., Momin, A. C., and Rege, S. G., 1959, An X-ray study of leucoxene from Quilon, India: Economic Geology, v. 54, p. 913–918.

Kildal, E. S., 1970, Geologisk karte over Norge, berggrunnskart, Maloy sheet: Norges Geologiske Undersokelse (Norway), scale 1:250,000.

Kinney, D. M., 1949, The Magnet Cove Rutile Company mine, Hot Springs County, Arkansas: U.S. Geological Survey Open-File Report 19, 12 p.

Klugman, M. A., 1966, Resume of the geology of the Laramie Anorthosite mass: Mountain Geologist, v. 3, p. 75–84.

Kolker, A., 1982, Mineralogy and geochemistry of Fe-Ti oxide and apatite (nelsonite) deposits and evaluation of the liquid immiscibility hypothesis: Economic Geology, v. 77, p. 1146–1158.

Komar, P. D., and Wang, C., 1984, Processes of selective grain transport and the formation of placers on beaches: Journal of Geology, v. 92, p. 637–655.

Korneliussen, A., 1980, Rutil: eklogittiske bergarter: Sunnfjord: Norges Geologiske Undersokelse Rapport 1717/5, 29 p.

Korneliussen, A., and Foslie, G., 1986, Rutile-bearing eclogites in the Sunnfjord region of western Norway: Norges Geologiske Undersokelse Bulletin 402, p. 65–71.

Korneliussen, A., Geis, H.-P., Gierth, E., Krause, H., Robins, B., and Schott, W., 1986, Titanium ores; An introduction to a review of titaniferous magnetite, ilmenite, and rutile deposits in Norway: Norges Geologiske Undersokelse Bulletin 402, p. 7–24.

Krause, H., and Pape, H., 1975, Mikroskopische untersuchungen der mineral-vergesellschaftung in erz und nebengestein der ilmenitlagerstätte Storgangen (süd-Norwegen): Norsk Geologisk Tidsskrift, v. 55, p. 387–422.

—— , 1977, Untersuchungen zum geologischen und petrographischen aufbau des Storgangen-ilmenitskörpers und seiner nebengesteinseinheiten (süd-Norwegen): Norsk Geologisk Tidsskrift, v. 57, p. 263–284.

Krause, H., and Zeino-Mahmalat, R., 1970, Untersuchungen an erz und neben-gesteinen der grube Blåfjell in SW-Norwegen: Norsk Geologisk Tidsskrift, v. 50, p. 45–88.

Krause, H., Gierth, E., and Schott, W., 1986, Fe-Ti deposits in the South Roga-land igneous complex, with special reference to the Åna-Sira anorthosite massif: Norges Geologiske Undersokelse Bulletin 402, p. 25–38.

Krogh, E. J., 1980, Geochemistry and petrology of glaucophane-bearing eclogites and associated rocks from Sunnfjord, western Norway: Lithos, v. 13, p. 355–380.

Kudrass, H.-R., 1982, Cores of Holocene and Pleistocene sediments from the east Australian continental shelf (SO-15 cruise 1980): Geologische Jahrbuch, v. D 56, p. 137–163.

Lang, H. D., 1970, Sekundare rutil-lagerstatten in Sierra Leone (West-Africa): Erzmetall, v. 23, p. 179–183.

Large, P. R., 1972, Metasomatism and scheelite mineralization at Bold Head, King Island: Australasian Institute of Mining and Metallurgy Proceedings no. 238, p. 31–45.

La Roche, H., de Kern, M., and Bolfa, J., 1962, Contribution a l'etude de l'alteration des ilmenites: Sciences de la Terre, v. 8, p. 215–248.

Larsen, E. S., Jr., 1942, Alkalic rocks of Iron Hill, Gunnison County, Colorado: U.S. Geological Survey Professional Paper 197-A, 64 p.

Lawrence, L. J., and Savage, E. N., 1975, Mineralogy of the titaniferous porphyry copper deposits of Melanesia: Australasian Institute of Mining and Metal-lurgy Proceedings, no. 256, p. 1–14.

Layton, W., 1966, Prospects of offshore mineral deposits on the eastern seaboard of Australia: Mining Magazine, v. 115, p. 344–351.

Lee, D. E., and Dodge, F.C.W., 1964, Accessory minerals in some granitic rocks in California and Nevada as a function of calcium content: American Miner-alogist, v. 79, p. 1660.

Lindberg, P. A., 1986, Fe-Ti-P mineralizations in the larvikite-lardalite complex, Oslo Rift: Norges Geologiske Undersokelse Bulletin 402, p. 93–98.

Lissiman, J. C., and Oxenford, R. J., 1973, The Allied Minerals N. L. heavy minerals sand deposit at Eneabba, Western Australia: Australasian Institute of Mining and Metallurgy (Parkville, Victoria), Western Australia confer-ence, preprint.

—— , 1975, Eneabba rutile-zircon-ilmenite sand deposits, W. A., *in* Knight, C. L., ed., Economic geology of Australia and Papua New Guinea: Aus-traliasian Institute of Mining and Metallurgy Monograph 5, p. 1062–1069.

Lister, G. F., 1966, The composition and origin of selected iron-titanium deposits: Economic Geology, v. 61, p. 275–310.

Llewellyn, T. O., and Sullivan, G. V., 1980, Recovery of rutile from a porphyry copper tailings sample: U.S. Bureau of Mines Report of Investigation 8462, 18 p.

Loughman, F. C., 1969, Chemical weathering of the silicate minerals: New York, American Elsevier, 154 p.

Lowrie, D. C., Horwitz, R. C., and Gemuts, I., 1967, Busselton and Augusta sheets: Geological Survey of Western Australia, scale 1:250,000.

Lowrie, D. C., Low, G. H., Playford, P. E., and Dedman, R., 1973, Dongara and Hill River sheets: Geological Survey of Western Australia, scale 1:250,000.

Lowrie, D. C., Wilde, S. A., and Walker, I. W., 1983, Collie sheet: Geological Survey of Western Australia, scale 1:250,000.

Lowright, R., Williams, E. G., and Dachille, F., 1972, An analysis of factors controlling deviations in the hydraulic equivalence in some modern sands: Journal of Sedimentary Petrology, v. 42, p. 635–645.

Lynd, L. E., 1980, Study of the mechanism and rate of ilmenite weathering: American Institute of Mining, Metallurgical, and Petroleum Engineers Transactions, v. 217, p. 311–318.

—— , 1985, Titanium, *in* Mineral facts and problems, 1985, ed.: U.S. Bureau of Mines Bulletin 675, p. 859–880.

—— , 1988, "Ilmenite," "Rutile," and "Titanium and titanium dioxide," *in* Mineral commodity summaries 1988: U.S. Bureau of Mines, p. 72–73, 132–133, 170–171.

Lynd, L. E., Sigurdson, H., North, C. H., and Anderson, W. W., 1954, Character-istics of titaniferous concentrates: American Institute of Mining, Metallurgi-cal, and Petroleum Engineers Transactions, v. 6, p. 817–824.

Macdonald, E. H., 1971a, Country report; Indonesia: United Nations Economic Commission for Asia and Far East Technical Bulletin, v. 5, p. 48–53.

—— , 1971b, Country report; West Mayalsia: United Nations Economic Com-mission for Asia and Far East Technical Bulletin, v. 5, p. 74–78.

—— , 1983, Alluvial mining: London, Chapman and Hall, 508 p.

Mader, D., 1980, Authigener rutil im Buntsandstein des Westeifel: Neues Jah-rbuch für Mineralogie Monatshefte, no. 3, p. 97–108.

Mallik, T. K., 1986, Micromorphology of some placer minerals from Kerala beach, India: Marine Geology, v. 71, p. 371–381.

Mallik, T. K., Vasudevan, V., Verghese, P. A., and Machado, T., 1987, The black sand placer deposits of Kerala Beach, southwest India: Marine Geology, v. 77, p. 129–150.

Mancini, A., Mancini, R., and Martinotti, G., 1979, Valorization of new titanium resource; Titaniferous eclogites: Proceedings IV-15, 10th World Mining Congress, Istanbul, 19 p.

Markewicz, F. J., 1969, Ilmenite deposits of New Jersey coastal plain, *in* Sub-itzky, S., ed., Geology of selected areas in New Jersey and eastern Pennsyl-vania: New Brunswick, New Jersey, Rutgers University Press, p. 363–382.

Marsh, S. P., 1979, Rutile mineralization in the White Mountain andalusite deposit, California: U.S. Geological Survey Open-File Report 79–1622, 7 p.

Marsh, S. P., and Sheridan, D. M., 1976, Rutile in Precambrian sillimanite-quartz gneiss and related rocks, east-central Front Range, Colorado: U.S. Geologi-cal Survey Professional Paper 959-G, 17 p.

Martens, J.H.C., 1935, Beach sands between Charleston, South Carolina, and Miami, Florida: Geological Society of America Bulletin, v. 46, p. 1563–1596.

Martinis, B., and Pasquare, G., 1971, Carta Geologica d'Italia, sheet 82, Genova: Servicio Geologico d'Italia, scale 1:100,000.

Mathis, J. M., and Sclar, C. B., 1980, The oxidation and titanium-enrichment mechanism of "altered ilmenite" grains in the Tertiary Kirkwood and Co-hansey Formations of New Jersey: Geological Society of America Abstracts with Programs, v. 12, p. 72.

Maud, R. R., 1968, Quaternary geomorphology and soil formation in coastal Natal: Zeitschrift Geomorphologie Supplementband 7, p. 155–199.

McCartan, L., Weems, R. E., and Lemon, E. M., 1990, Quaternary stratigraphy in the vicinity of Charleston, South Carolina, and its relationship to local seismicity and tectonics: U.S. Geological Survey Professional Paper 1367-A (in press).

McCauley, C. K., 1960, Exploration for heavy minerals on Hilton Head Island, South Carolina: South Carolina Division of Geology Bulletin 26, 13 p.

McElroy, C. T., 1962, The geology of the Clarence–Moreton Basin: Geological Survey of New South Wales Geological Memoir 9, p. 1–172.

McIntyre, A., and 7 others, 1976, Glacial North Atlantic 18,000 years ago; A CLIMAP reconstruction, *in* Clise, R. M., and Hays, J. P., eds., CLIMAP investigation of late Quaternary paleoceanography and paleoclimatology: Geological Society of America Memoir 145, p. 43–76.

McIntyre, D. D., 1959, The hydraulic equivalence and size distributions of some mineral grains from a beach: Journal of Geology, v. 67, p. 278–301.

McKellar, J. B., 1975, The eastern Australia rutile province, *in* Knight, C. L., ed., Economic geology of Australia and Papua New Guinea: Australasian Insti-tute of Mining and Metallurgy Monograph 5, p. 1055–1061.

McLaughlin, R.J.W., 1955, Geochemical changes due to weathering under vary-ing climatic conditions: Geochimica et Cosmochimica Acta, v. 8, p. 109–130.

Meilke, H., and Schreyer, W., 1972, Magnetite-rutile assemblages in metapelites of the Fichtelgebirge, Germany: Earth and Planetary Science Letters, v. 16, p. 423–428.

Melville, G., 1984, Headlands and offshore islands as dominant controlling factors during late Quaternary barrier formation in the Forster–Tuncurry area, New South Wales, Australia: Sedimentary Geology, v. 39, p. 243–271.

Mertie, J. B., Jr., 1979, Monazite in the granitic rocks of the southeastern Atlantic States; An example of the use of heavy minerals in geologic exploration: U.S. Geological Survey Professional Paper 1094, 79 p.

Miller, R. L., and Zeigler, J. M., 1958, A model relating sediment pattern in the region of shoaling waves, breaker zone, and foreshore: Journal of Geology, v. 66, p. 417–441.

Minard, J. P., Force, E. R., and Hayes, G. W., 1976, Alluvial ilmenite placer deposits, central Virginia: U.S. Geological Survey Professional Paper 959-H, 15 p.

Mineração Metalurgia, 1977, Projecto titânio: v. 41, p. 26–29.

Mitchell, R. H., 1986, Kimberlites: New York, Plenum Press, 442 p.

Mohr, D. W., and Newton, R. C., 1983, Kyanite-staurolite metamorphism in sulfidic schists of the Anakeesta Formation, Great Smoky Mountains, North Carolina: American Journal of Science, v. 283, p. 97–134.

Moore, W. J., and Czamanske, G. K., 1973, Compositions of biotites from unaltered and altered monzonitic rocks in the Bingham mining district, Utah: Economic Geology, v. 68, p. 269–274.

Morad, S., 1986, SEM study of authigenic rutile, anatase, and brookite in Proterozoic sandstones from Sweden: Sedimentary Geology, v. 46, p. 77–89.

Morad, S., and Aldahan, A. A., 1985, Leucoxene-calcite-quartz aggregates in sandstones and the relation to decomposition of sphene: Neues Jahrbook für Mineralogie Monatshefte no. 10, p. 458–468.

——— , 1986, Alteration of detrital Fe-Ti oxides in sedimentary rocks: Geological Society of America Bulletin, v. 97, p. 567–578.

Morley, I. W., 1981, Black sands; A history of the mineral sand mining industry in eastern Australia: Brisbane, University of Queensland Press, 278 p.

Morse, S. A., 1982, A partisan review of Proterozoic anorthosites: American Mineralogist, v. 67, p. 1087–1100.

Mukerjee, N. K., 1964, A study of the monazite-bearing ilmenite deposit from Bothaville, Orange Free State, Union of South Africa: New Delhi, 22nd International Geological Congress Report, sec. 16, p. 481–503.

Nafziger, R. H., and Elger, G. W., 1987, Preparation of titanium feedstock from Minnesota ilmenite by smelting and sulfation-leaching: U.S. Bureau of Mines Report of Investigation 9065, 13 p.

Naldrett, A. J., 1979, Partitioning of Fe, Co, Ni, and Cu between sulfide liquid and basaltic melts, and the composition of Ni-Cu sulfide deposits; A reply and further discussion: Economic Geology, v. 74, p. 1520–1528.

Narayanaswami, S., and Mahadevan, T. M., 1964, Geology of Trivandrum and Madras areas: 22nd International Geological Congress, New Delhi, Guide to excursion C-23, 13 p.

Nathan, H. D., 1969, The geology of a portion of the Duluth Complex, Cook County [Ph.D. thesis]: Minneapolis, University of Minnesota, 198 p.

National Materials Advisory Board, 1983, Titanium; Past, present, and future: U.S. National Academy Press Publication NMAB-392, 209 p.

Nedelcu, L., 1986, The significance of titanium minerals in some crystalline schists from Romania, in Mineral parageneses: Athens, Theophrastus, p. 623–645.

Neiheisel, J., 1958, Origin of the dune system on the Isle of Palms, S.C.: South Carolina Division of Geology Mineral Industries Laboratory Bulletin, v. 2, p. 46–51.

——— , 1962, Heavy-mineral investigations of Recent and Pleistocene sands of the lower coastal plain of Georgia: Geological Society of America Bulletin, v. 73, p. 365–374.

——— , 1976, Heavy minerals in aeroradioactive high areas of the Savannah River flood plain and deltaic plain: South Carolina Division of Geology Geologic Notes, v. 20, p. 45–51.

Nesbitt, B. E., and Kelly, W. C., 1980, Metamorphic zonation of sulfides, oxides, and graphite in and around the orebodies at Ducktown, Tennessee: Economic Geology, v. 75, p. 1010–1021.

Newhouse, W. H., and Hagner, A. F., 1951, Preliminary report on the titaniferous iron deposits of the Laramie Range, Wyoming: U.S. Geological Survey Open-File Report, 17 p.

——— , 1957, Geologic map of anorthosite areas, southern part of Laramie Range, Wyoming: U.S. Geological Survey Map MF-119, scale 1:63,360.

Nielsen, N.-A., 1972, The ilmenomagnetite-apatite deposit at Kodal, S. Norway: Industrial Minerals, April, p. 35–37.

Oakeshott, G. B., 1958, Geology and mineral deposits of San Fernando Quadrangle, Los Angeles County, California: California Division of Mines Bulletin 172, 147 p.

Oaks, R. Q., Coch, N. K., Sanders, J. E., and Flint, R. F., 1974, Post-Miocene shorelines and sea levels, southeastern Virginia, in Oaks, R. Q., and Du Bar, J. R., eds., Post-Miocene stratigraphy, central and southern Atlantic coastal plain: Logan, Utah State University, p. 53–87.

Oeschger, H.,1985, The contribution of ice core studies to the understanding of environmental processes: American Geophysical Union Geophysical Monograph 33, p. 9–17.

Oftedahl, C., 1960, Permian rocks and structures of the Oslo region: Norges Geologiske Undersokelse (Bulletin) 208, p. 298–343.

Opdike, N. D., Spangler, D. P., Smith, D. L., Jones, D. S., and Lindquist, R. C., 1984, Origin of the epeirogenic uplift of Pliocene–Pleistocene beach ridges in Florida, and development of the Florida karst: Geology, v. 12, p. 226–228.

Ortega-Gutierez, F., 1981, Metamorphic belts of southern Mexico and their tectonic significance: Geofisica International, v. 20-3, p. 177–202.

Overstreet, W. C., Yates, R. G., and Griffitts, W. R., 1963, Heavy minerals in the saprolite of the crystalline rocks in the Shelby Quadrangle, North Carolina: U.S. Geological Survey Bulletin 1162-F, 31 p.

Owens, J. P., 1985, Sediment dispersal pattern in the emerged Atlantic coastal plain (New Jersey–Georgia) from Early Cretaceous to Holocene; A guide to potential placer deposits [abs.]: U.S. Geological Survey Circular 949, p. 39–40.

Pasteels, P., Demaiffe, D., and Michot, J., 1979, U-Pb and Rb-Sr geochemistry of the eastern part of the south Rogaland igneous complex, southern Norway: Lithos, v. 12, p. 199–208.

Pasteris, J. D., 1980, The significance of groundmass ilmenite and megacryst ilmenite in kimberlites: Contributions to Mineralogy and Petrology, v. 75, p. 315–325.

——— , 1985, Relationships between temperature and oxygen fugacity among Fe-Ti oxides in two regions of the Duluth Complex: Canadian Mineralogist, v. 23, p. 111–127.

Patterson, S. H., Kurtz, H. F., Olson, J. C., and Neeley, C. L., 1986, World bauxite resources: U.S. Geological Survey Professional Paper 1076-B, 151 p.

Paulson, E. G., 1964, Mineralogy and origin of the titaniferous deposit at Pluma Hidalgo, Oaxaca, Mexico: Economic Geology, v. 59, p. 753–767.

Peers, R., 1975a, Leeuwin block, in Geology of Western Australia: Geological Survey of Western Australia Memoir 2, p. 102–104.

——— , 1975b, Northampton block, in Geology of Western Australia: Geological Survey of Western Australia Memoir 2, p. 104–106.

Petersen, J. S., 1978, Structure of the larvikite-lardalite complex, Oslo region, Norway, and its evolution: Geologische Rundschau, v. 67, p. 330–342.

Peterson, C. D., Komar, P. D., and Scheidegger, K. F.,1985, Distribution, geometry, and origin of heavy mineral placer deposits on Oregon beaches: Journal of Sedimentary Petrology, v. 56, p. 67–77.

Pettijohn, F. J., 1957, Sedimentary rocks: New York, Harper and Row, 718 p.

Philpotts, A. R., 1967, Origin of certain iron-titanium oxide and apatite rocks: Economic Geology, v. 62, p. 303–315.

Pinnell, D. B., and Marsh, J. A., 1954, Summary geological report of the titaniferous iron deposits of the Laramie Range, Albany County: Wyoming Mines Magazine, v. 44, 30 p.

Pirkle, E. C., and Yoho, W. H., 1970, The heavy mineral ore body of Trail Ridge, Florida: Economic Geology, v. 65, p. 17–30.

Pirkle, E. C., Yoho, W. H., and Hendry, C. W., 1970, Ancient sea level stands in Florida: Florida Bureau of Geology Geological Bulletin 52, 61 p.

——— , 1971, North Florida setting for 15th field conference, Southeastern Geological society, *in* Geological review of some north Florida mineral resources: Tallahassee, Southeastern Geological Society, p. 1–15.

Pirkle, E. C., Pirkle, W. A., and Yoho, W. H., 1974, The Green Cove Springs and Boulougne heavy-mineral sand deposits of Florida: Economic Geology, v. 67, p. 1129–1137.

——, 1977, The Highland heavy-mineral sand deposit on Trail Ridge in northern peninsular Florida: Florida Bureau of Geology Report of Investigation 84, 50 p.

Pirkle, E. C., Pirkle, F. L., Pirkle, W. A., and Stayert, P. R., 1984, The Yulee heavy mineral sand deposits of northeastern Florida: Economic Geology, v. 79, p. 725–737.

Pirkle, F. L., 1975, Evaluation of possible source regions of Trail Ridge sands: Southeastern Geology, v. 17, p. 93–114.

Pirkle, F. L., and Czel, L. J., 1983, Marine fossils from region of Trail Ridge, a Georgia–Florida landform: Southeastern Geology, v. 24, p. 31–38.

Poulose, K. V., 1972, Heavy mineral and glass sand deposits of Kerala coast: Indian Minerals, v. 26, p. 118–124.

Prakash, T. N., and Verghese, P. A., 1987, Seasonal beach changes along Quilon district coast: Journal of the Geological Society of India, v. 29, p. 390–398.

Pryor, W. A., 1960, Cretaceous sedimentation in upper Mississippi embayment: American Association of Petroleum Geologists Bulletin, v. 44, p. 1473–1504.

Puffer, J. H., and Cousminer, H. L., 1982, Factors controlling the accumulation of titanium-iron oxide-rich sands in the Cohansey Formation, Lakehurst area, New Jersey: Economic Geology, v. 77, p. 379–391.

Purucker, M., 1983, Time of formation of soft iron ore on the Gunflint and Mesabi ranges: Economic Geology, v. 78, p. 502–506.

Quirk, R., and Eilertson, N. A., 1963, Methods and costs of exploration and pilot plant testing of ilmenite-bearing sands, Lakehurst mine, The Glidden Co., Ocean County, N.J.: U.S. Bureau of Mines Information Circular 8197, 64 p.

Rachele, L. D., 1976, Palynology of the Legler Lignite; A deposit in the Tertiary Cohansey Formation of New Jersey, U.S.A.: Reviews in Paleobotany and Palynology, v. 22, p. 225–252.

Raedeke, L. D., 1982, Mineralogy and petrology of layered intrusions; A review: Lunar and Planetary Institute Report 82-01, p. 128–134.

Ramberg, H., 1948, Titanic iron ore formed by dissociation of silicates in granulite facies [Greenland]: Economic Geology, v. 43, p. 553–569.

——, 1952, The origin of metamorphic and metasomatic rocks: Chicago, Illinois, University of Chicago Press, 317 p.

Rampacek, C., 1982, An overview of mining and mineral processing waste as a resource: Resources and Conservation, v. 9, p. 75–86.

Raufuss, W., 1973, Structur, schwermineralführung, genese, and bergbau der sedimentären Rutil Lagerstätten in Sierra Leone (Westafrika): Geologische Jahrbuch Reihe D, no. 5, p. 3–52.

Reed, D. F., 1949a, Investigation of Christy titanium deposit, Hot Spring County, Ark.: U.S. Bureau of Mines Report of Investigation 4592, 10 p.

——, 1949b, Investigation of Magnet Cove rutile deposit, Hot Spring County, Arkansas: U.S. Bureau of Mines Report of Investigation 4593, 9 p.

Reich, V., Kudrass, H.-R., and Wiedicke, M., 1982, Heavy minerals of the east Australian shelf sediments between Newcastle and Fraser Island: Geologische Jahrbuch, v. D 56, p. 179–195.

Reynolds, R. L., and Goldhaber, M. B., 1978, Origin of south Texas roll-type uranium deposits; 1, Alteration of iron-titanium oxide minerals: Economic Geology, v. 73, p. 1677–1689.

Rich, F. L., 1985, Palynology and paleoecology of a lignitic peat from Trail Ridge, Florida: Florida Bureau of Geology Information Circular 100, 15 p.

Riezebos, P. A., 1979, Compositional downstream variation of opaque and translucent heavy residues in some modern Rio Magdalena sands (Columbia): Sedimentary Geology, v. 24, p. 197–225.

Rittenhouse, G., 1943, Transportation and deposition of heavy minerals: Geological Society of America Bulletin, v. 54, p. 1725–1780.

Robinson, P., and Tracy, R. J., 1977, Sulfide-silicate-oxide equilibria in sillimanite-K feldspar grade pelitic schists [abs.]: EOS Transactions of the American Geophysical Union, v. 58, p. 524.

Robinson, P., Hollocher, K. J., Tracy, R. J., and Dietsch, C. J., 1982, High grade regional metamorphism in south-central Massachusetts: Connecticut Geological and Natural History Survey Guidebook 5, p. 289–340.

Robson, D. F., and Sampath, N., 1977, Geophysical response of heavy-mineral sand deposits at Jerusalem Creek, New South Wales: (Australia) Bureau of Mineral Resources Journal of Geology and Geophysics, v. 2, p. 149–154.

Roehler, H. W., 1989, Origin and distribution of six heavy mineral placer deposits in coastal-marine sandstones in the upper Cretaceous McCourt Sandstone tongue of the Rock Springs Formation, southwest Wyoming: U.S. Geological Survey Bulletin 1867, 34 p.

Romey, W. D., 1968, An evaluation of some differences between anorthosites in massifs and in layered complexes: Lithos, v. 1, p. 230–241.

Rose, C. K., and Shannon, S. S., Jr., 1960, Cebolla Creek titaniferous iron deposits, Gunnison County, Colorado: U.S. Bureau of Mines Report of Investigation 5679, 30 p.

Rose, E. R., 1969, Geology of titanium and titaniferous deposits of Canada: Canada Geological Survey Economic Geology Report 25, 177 p.

Ross, C. S., 1941, Occurrence and origin of the titanium deposits of Nelson and Amherst Counties, Virginia: U.S. Geological Survey Professional Paper 198, 59 p.

Roy, P. S., 1977, Does the Hunter River supply sand to the New South Wales coast today? Journal of the Royal Society of New South Wales, v. 110, p. 17–24.

——, 1982, Regional geology of the central and northern New South Wales coast: Geologische Jahrbuch, v. D 56, p. 25–35.

Roy, P. S., and Crawford, E. A., 1977, Significance of sediment distribution in major coastal rivers, northern New South Wales: 3rd Australian Conference on Coastal and Ocean Engineering, Institution of Engineers, Melbourne, p. 177–184.

Roy, P. S., and Thom, B. G., 1981, Late Quaternary marine deposition in New South Wales and southern Queensland; An evolutionary model: Geological Society of Australia Journal, v. 28, p. 471–489.

Rubey, W. W., 1933, The size distribution of heavy minerals within a waterlaid sandstone: Journal of Sedimentary Petrology, v. 3, p. 3–29.

Ruddiman, W. F., 1985, Climate studies in ocean cores, *in* Hecht, A. D., ed., Paleoclimate analysis and modeling: New York, Wiley-Interscience, p. 197–253.

Rumble, D., III, 1973, Fe-Ti oxide minerals from regionally metamorphosed quartzites of western New Hampshire: Contributions to Mineralogy and Petrology, v. 42, p. 181–195.

——, 1976, Oxide minerals in metamorphic rocks, *in* Rumble, D., III, ed., Oxide minerals: Mineralogical Society of America Short Course Notes, v. 3, p. R1–24.

Russell, E. E., 1967, Mineral resources summary of the Stantonville Quadrangle, Tennessee: Tennessee Division of Geology of Mineral Resources Summary MRS13-NW (and map), 18 p.

——, 1975, Stratigraphy of the outcropping upper Cretaceous in western Tennessee: Tennessee Division of Geology Bulletin 75, p. A1–A65.

Russell, R. D., 1937, Mineral composition of Mississippi River sands: Geological Society of America Bulletin, v. 48, p. 1307–1348.

Rust, B. R., and Jones, B. G., 1987, The Hawkesbury Sandstone south of Sydney, Australia; Triassic analogue for the deposit of a large braided river: Journal of Sedimentary Petrology, v. 57, p. 222–233.

Saggerson, E. P., and Turner, L. M., 1978, Metamorphic map of Africa: UNESCO Commission for the Geological Map of the World, scale 1:10,000,000.

Sainsbury, C. L., 1968, Tin and beryllium deposits of the central York Mountains, western Seward Peninsula, Alaska, *in* Ridge, J. D., ed., Ore deposits of the United States, 1933–1967, v. 2: New York, American Institute of Mining, Metallurgical, and Petroleum Engineers, p. 1555–1572.

Sallenger, A. H., 1979, Inverse grading and hydraulic equivalence in grain-flow deposits: Journal of Sedimentary Petrology, v. 49, p. 553–562.

Schluter, H. U., 1982, Results of a reflection seismic survey in shallow water areas off east Australia, Yamba to Tweed Heads: Geologische Jahrbuch, v. D 56, p. 77–95.

Schmidt, R. G., 1985, High-alumina hydrothermal systems in volcanic rocks and their significance to mineral prospecting in the Carolina slate belt: U.S. Geological Survey Bulletin 1562, 59 p.

Scholle, P. A., 1979, Constituents, textures, cements, and porosities of sandstones and associated rocks: American Association of Petroleum Geologists Memoir 28, 201 p.

Scholtz, D. L., 1936, The magmatic nickeliferous ore deposits of East Griqualand and Pondoland: Transactions of the Geological Society of South Africa, v. 39, p. 81–210.

Schumm, S. A., Mosley, M. P., and Weaver, W. E., 1987, Experimental fluvial geomorphology: New York, Wiley-Interscience, 413 p.

Scull, B. J., 1958, Origin and occurrence of barite in Arkansas: Arkansas Geological and Conservation Commission Information Circular 18, 101 p.

Seki, Y., Onuki, H., Oba, T., and Mori, R., 1971, Sambagawa metamorphism in the central Kii Peninsula, Japan: Japanese Journal of Geology and Geography, v. 41, p. 65–78.

Semeniuk, V., and Johnson, D. P., 1982, Recent and Pleistocene beach/dune sequences, Western Australia: Sedimentary Geology, v. 32, p. 301–328.

Semeniuk, V., and Searle, D. J., 1986, Variability of Holocene sealevel history along the southwestern coast of Australia; Evidence for the effect of significant local tectonism: Marine Geology, v. 72, p. 47–58.

Shepherd, M. S., 1986, Australian heavy mineral reserves and world trends, *in* Australia; A world source of ilmenite, rutile, monazite, and zircon: Perth, Australasian Institute of Mining and Metallurgy, p. 61–68.

—— , 1990, Eneabba heavy mineral sand placers, Western Australia, *in* Geology and Mineral Deposits of Australia and Papua New Guinea: Parkville, Australasian Institute of Mining and Technology (in press).

Sheridan, D. M., Taylor, R. B., and Marsh, S. P., 1968, Rutile and topaz in Precambrian gneiss, Jefferson and Clear Creek Counties, Colorado: U.S. Geological Survey Circular 567, 7 p.

Sherman, G. D., 1952, The titanium content of Hawaiian soils and its significance: Soil Science Society of America Proceedings, v. 16, p. 15–18.

Shideler, G. L., and Smith, K. P., 1984, Regional variability of beach and fore-dune characteristics along the Texas Gulf Coast barrier system: Journal of Sedimentary Petrology, v. 54, p. 507–526.

Short, A. D., 1987, Modes, timing, and volume of Holocene cross-shore and aeolian sand transport, southern Australia, *in* Coastal sediments '87: New York, American Society of Civil Engineers, v. 2, p. 1925–1937.

—— , 1988, Holocene coastal dune formation in southern Australia; A case study: Sedimentary Geology, v. 55, p. 121–142.

Short, N. M., 1961, Geochemical variations in four residual soils: Journal of Geology, v. 69, p. 534–571.

Siever, R., and Woodford, N., 1979, Dissolution kinetics and the weathering of mafic minerals: Geochimica et Cosmochimica Acta, v. 43, p. 717–724.

Sillitoe, R. H., 1983, Unconventional metals in porphyry deposits, *in* Shanks, W. C. III, ed., Cameron volume on unconventional mineral deposits: New York, Society of Mining Engineers, p. 207–221.

Sinha, R. K., 1967, A treatise on industrial minerals of India: Bombay, Allied Publishers, 513 p.

Skjerlie, F. J., and Pringle, I. R., 1978, A Rb/Sr whole-rock isochron date from the lowermost gneiss complex of the Gaular area, west Norway and its regional implications: Norsk Geologisk Tidsskrift, v. 58, p. 259–265.

Slingerland, R. L., 1977, The effects of entrainment on the hydraulic equivalence relationships of light and heavy minerals in sands: Journal of Sedimentary Petrology, v. 47, p. 753–770.

—— , 1984, Role of hydraulic sorting in the origin of fluvial placers: Journal of Sedimentary Petrology, v. 54, p. 137–150.

Slingerland, R., and Smith, N. D., 1986, Occurrence and formation of water-land placers: Annual Review of Earth and Planetary Science, v. 14, p. 113–147.

Smith, N. D., and Minter, W.E.L., 1980, Sedimentological control of gold and uranium in two Witwatersrand placers: Economic Geology, v. 75, p. 1–14.

Smithson, S. B., and Hodge, D. S., 1972, Field relations and gravity interpretation in the Laramie Anorthosite complex: Laramie, University of Wyoming Contributions to Geology, v. 11, p. 43–59.

Soman, K., 1985, Origin and geologic significance of the Chavara placer deposit, Kerala: Current Science, v. 54, p. 280–281.

Sorensen, H., 1974, Alkali syenites, feldspathoidal syenites, and related lavas, *in* Sorensen, H., ed., The alkaline rocks: London, Wiley, p. 22–52.

Southwick, D. L., 1968, Mineralogy of a rutile- and apatite-bearing ultramafic chlorite rock, Harford County, Maryland: U.S. Geological Survey Professional Paper 600-C, p. 38–44.

Spear, F. S., 1981, An experimental study of hornblende stability and compositional variability in amphibolite: American Journal of Science, v. 281, p. 697–734.

Spencer, R. V., 1948, Titanium minerals in Trail Ridge, Florida: U.S. Bureau of Mines Report of Investigation 4208, 21 p.

Spencer, R. V., and Williams, F. R., 1967, The development of a rutile mining industry in Sierra Leone: Proceedings—General, 8th Commonwealth Mining and Metallurgical Congress, Australia and New Zealand, 1965, Institute of Mining and Metallurgy, London, v. 6, p. 1337–1341.

Spencer, S., 1964, African creeks I have been up: New York, McKay, 236 p.

Starmer, I. C., 1985a, The evolution of the south Norwegian Proterozoic as revealed by the major and mega-tectonics of the Kongsberg and Bamble sectors, *in* Tobi, A. C., and Touret, J.L.R., eds., The deep Proterozoic crust in the North Atlantic provinces: Hingham, Massachusetts, Reidel, p. 259–290.

—— , 1985b, Geological map of the Bamble sector, south Norway, 3 sheets: NATO Advanced Studies Institute Excursion guide, scale 1:100,000.

Stephens, A. W., 1982a, Quaternary coastal sediments of the southeast Queensland: Geologische Jahrbuch, v. D 56, p. 37–47.

—— , 1982b, Surficial sediments of the southern Queensland shelf; Southport–Point Lookout and Fraser Island areas: Geologische Jahrbuch, v. D 56, p. 125–135.

Stephenson, R. C., 1945, Titaniferous magnetite deposits of the Lake Sanford area, New York: New York State Museum Bulletin 340, 78 p.

Stone, B. D., and Force, E. R., 1980, The Port Leyden, New York, heavy mineral deposit: New York State Museum Bulletin 436, p. 57–61.

—— , 1983, Coarse-grained glaciolacustrine deltaic sedimentation in ice-marginal and ice-channel environments: Geological Society of America Abstracts with programs, v. 15, p. 126.

Storch, R. H., and Holt, D. C., 1963, Titanium placer deposits of Idaho: U.S. Bureau of Mines Report of Investigation 6319, 69 p.

Subrahmanyam, N. P., and Rao, G.V.U., 1980, Niobium and tantalum in the rutile of Karala beach sands: Geological Society of India Journal, v. 21, p. 623–626.

Subrahmanyam, N. P., and 5 others, 1982, Alteration of beach sand ilmenite from Manavalakurichi, Tamil Nadu, India: Geological Society of India Journal, v. 23, p. 168–174.

Sullivan, G. V., and Llewellyn, T. O., 1981, Occurrence and recovery of rutile from western copper mill tailings: Society of Mining Engineers Preprint 81-333, 7 p.

Sundquist, E. T., and Broecker, W. S., 1985, The carbon cycle and atmospheric CO_2; Natural variations, Archean to present: American Geophysical Union Geophysical Monograph 32, 627 p.

Swift, D.J.P., 1968, Coastal erosion and transgressive stratigraphy: Journal of Geology, v. 76, p. 444–456.

Taylor, H. P., Jr., and Coleman, R. G., 1968, O^{18}/O^{16} ratios of coexisting minerals in glaucophane-bearing metamorphic rocks: Geological Society of America Bulletin, v. 79, p. 1727–1756.

Temple, A. K., 1966, The alteration of ilmenite: Economic Geology, v. 61, p. 695–714.

Temple, A. K., and Grogan, R. M., 1965, Carbonatite and related alkalic rocks at Powderhorn, Colorado: Economic Geology, v. 60, p. 672–692.

Teufer, G., and Temple, A. K., 1966, Pseudorutile; A new mineral intermediate between ilmenite and rutile in the natural alteration of ilmenite: Nature, v. 211, p. 179–181.

Thom, B. G., 1983, Transgressive and regressive stratigraphies of coastal sand bodies in southeast Australia: Marine Geology, v. 56, p. 137–158.

Thom, B. G., and Roy, P. S., 1985, Relative sea levels and coastal sedimentation in southeastern Australia in the Holocene: Journal of Sedimentary Petrology, v. 55, p. 257.

Thom, B. G., Adams, R. P., Cazeau, C. J., and Heron, S. D., 1972, Aspects of the texture and mineralogy of surficial sediments, Horry and Marion Counties, South Carolina: Southeastern Geology, v. 14, p. 39–58.

Thom, B. G., Bowman, G. M., and Roy, P. S., 1981, Late Quaternary evolution of coastal sand barriers, Port Stephens–Myall Lakes area, central New South Wales, Australia: Quaternary Research, v. 15, p. 345–364.

Thomas, B. I., and Berryhill, R. V., 1962, Reconnaissance studies of Alaskan beach sands, eastern Gulf of Alaska: U.S. Bureau of Mines Report of Investigation 5986, 40 p.

Thompson, C. H., and Bowman, G. M., 1984, Subaerial denudation and weathering of vegetated coastal dunes in eastern Australia, *in* Thom, B. G., ed., Coastal geomorphology in Australia: Sydney, Academic Press, p. 263–290.

Thompson, C. H., and Ward, W. T., 1975, Soil landscapes of North Stradbroke Island: Proceedings of the Royal Society of Queensland, v. 86, p. 9–14.

Thompson, J. V., 1987, Titanium resource in Colorado equals all other U.S. deposits: Engineering and Mining Journal, v. 188, p. 27–30.

Thorman, C. H., 1977, Geologic map of the Monrovia Quadrangle, Liberia: U.S. Geologial Survey Map I-775-D, scale 1:250,000.

Tipper, G. H., 1914, The monazite sands of Travancore: Geological Survey of India Records, v. 44, p. 186–196.

Toewe, E. C., Schatz, R. W., Carmichael, R. L., Gonser, B. W., and Goldberger, W. M., 1971, Evaluation of columbian-bearing rutile deposits, Magnet Cove, Arkansas: U.S. Department of the Interior Contract Report 14-01-001-1738, 134 p.

Tourtelot, H. A., 1968, Hydraulic equivalence of grains of quartz and heavier minerals, and implications for the study of placers: U.S. Geological Survey Professional Paper 594-F, p. F1–F13.

—— , 1983, Continental aluminous weathering sequences and their climatic implications in the United States: U.S. Geological Survey Circular 822, p. 1–5.

Towner, R. R., Gray, J. M., and Porter, L. M., 1988, International strategic minerals inventory summary report; Titanium: U.S. Geological Survey Circular 930-G, 58 p.

Troger, W. E., 1928, Alkaligesteine aus der Sierra do Salitre in westlichen Minas Gerais, Brazilien: Zentralblatt für Mineralogie, Geologie, and Paleontologie, 1928A, p. 202–204.

—— , 1935, Spezielle Petrographie der Eruptivgesteine: Berlin, Verlag Deutsch Mineralogischen Gesellschaft, Berlin, p. 279–282.

Turekian, K. K., 1977, Geochemical distribution of elements, *in* McGraw-Hill encyclopedia of science of technology: New York, McGraw-Hill, p. 627, 630.

Turner, F. J., 1968, Metamorphic petrology; Mineralogical and field aspects: New York, McGraw-Hill, 403 p.

Turner, R., 1986, Brazilian titanium: Engineering and Mining Journal, v. 187, p. 40–42.

Udubasa, G., 1982, Rutile of post-magmatic mineral formation, *in* Ore genesis; G. C. Amstutz, ed., The state of the art: Berlin, Springer-Verlag, p. 784–793.

Ulbrich, H.H.G.J., and Gomes, C. B., 1981, Alkaline rocks from continental Brazil: Earth Science Reviews, v. 17, p. 135–154.

United States Bureau of Mines, 1986, Domestic consumption trends, 1972–82, and forecasts to 1993 for twelve major metals: U.S. Bureau of Mines Open-File Report 27–86, 156 p.

Ussing, N. V., 1911, Geology of the country around Julianehaab, Greenland: Meddelelserom Gronland, v. 38, p. 1–376.

Valeton, I., 1972, Bauxites: New York, American Elsevier, 226 p.

van Andel, Tj. H., 1950, Provenance, transport, and deposition of Rhine sediments: Wageningen, Veenman und Zonen, 129 p.

Vos, R. G., and Hobday, D. K., 1977, Storm beach deposits in the late Paleozoic Ecca Group of South Africa: Sedimentary Geology, v. 19, p. 217–232.

Wahlstrom, E. E., 1948, Pre-Fountain and Recent weathering on Flagstaff Mountain near Boulder, Colorado: Geological Society of America Bulletin, v. 59, p. 1173–1190.

Walker, T. R., 1967, Formation of red beds in modern and ancient deserts: Geological Society of America Bulletin, v. 78, p. 353–368.

Wall Street Journal, 1976, Buttes says studies on Colorado prospect indicate major deposit of titanium ore: Wall Street Journal, v. 187, no. 37, p. 6 (Feb. 24).

Ward, W. T., 1977, Sand movement on Fraser Island; A response to changing climates: University of Queensland Department of Anthropology Papers, v. 8, p. 113–126.

—— , 1978, Notes on the origin of Stradbroke Island: University of Queensland Department of Geology Papers, v. 8, p. 97–104.

Ward, W. T., Little, I. P., and Thompson, C. H., 1979, Stratigraphy of two sandrocks at Rainbow Beach, Queensland, Australia, and a note on humate composition: Palaeogeography, Palaeoclimatology, and Palaeoecology, v. 26, p. 305–316.

Watson, T. L., 1912, Kragerite, a rutile-bearing rock from Kragerø, Norway: American Journal of Science, v. 34, p. 509–514.

Watson, T. L., and Taber, S., 1913, Geology of the titanium and apatite deposits of Virginia: Virginia Geological Survey Bulletin 3A, 308 p.

Weiblen, P. W., and Morey, G. B., 1980, A summary of the stratigraphy, petrology, and structure of the Duluth Complex: American Journal of Science, v. 280A, p. 88–133.

Welch, B. K., 1964, The ilmenite deposits of Geographe Bay: Australasian Institute of Mining and Metallurgy Proceedings, no. 211, p. 25–48.

Welch, B. K., Sofoulis, J., and Fitzgerald, A.C.F., 1975, Mineral sand deposits of the Capel area, West Australia, *in* Knight, C. L., ed., Economic geology of Australia and Papua New Guinea: Australasian Institute of Mining and Metallurgy Monograph 5, p. 1070–1087.

Wells, N., 1960, Total elements in topsoils from igneous rocks; An extension of geochemistry: Journal of Soil Science, v. 11, p. 409–424.

White, J. R., and Williams, E. G., 1967, The nature of a fluvial process as defined by settling velocities of heavy and light minerals: Journal of Sedimentary Petrology, v. 37, p. 530–539.

Whitworth, H. F., 1959, The zircon-rutile deposits on the beaches of the east coast of Australia with special reference to their mode of occurrence and the origin of the minerals: New South Wales Department of Mines Technical Report 4, p. 7–60.

Wiedner, J. R., 1982, Iron-oxide magmas in the system Fe-C-O: Canadian Mineralogist, v. 20, p. 555–566.

Wilcox, J. T., 1971, Preliminary investigation of heavy minerals in the McNairy sand of west Tennessee: Tennessee Division of Geology Report of Investigations 31, 11 p.

Williams, L., 1967, Heavy minerals in South Carolina: South Carolina Division of Geology Bulletin 35, 35 p.

Williams, S. A., and Cesbron, F. P., 1977, Rutile and apatite; Useful prospecting guides for porphyry copper deposits: Mineralogical Magazine, v. 41, p. 288–292.

Wilmart, E., and Duchesne, J. C., 1987, Geothermobarometry of igneous and metamorphic rocks around the Åna-Sira anorthosite massif; Implications for the depth of emplacement of the South Norwegian anorthosites: Norsk Geologisk Tidsskrift, v. 67, p. 185–196.

Wilson, A. F., 1964, The petrological features and structural setting of Australian granulites and charnockites: New Delhi, 22nd International Geological Congress Report, sec. 13, p. 21–44.

—— , 1969, Granulitic terrains and their tectonic setting and relationship to associated metamorphic rocks in Australia: Geological Society of Australia Special Publication 2, p. 243–258.

Winward, K., and Nicholson, D. A., 1974, Quaternary coastal sediments, *in* Markham, N. L., and Basden, H., eds., The mineral deposits of New South Wales: Geological Survey of New South Wales, p. 597–621.

Wise, W. S., 1975, The origin of the assemblage quartz + Al-silicate + rutile + Al-phosphate: Fortschrift Mineralogie, v. 52, p. 151–159.

—— , 1977, Mineralogy of the Champion mine, White Mountains, California: Mineralogical Record, Nov.-Dec., p. 478–486.

Wright, T. L., and Peck, D. L., 1978, Crystallization and differentiation of the

E. R. Force

Alae magma, Alae lava lake, Hawaii: U.S. Geological Survey Professional Paper 935 C, 20 p.

Wynn, J. C., and Grosz, A. E., 1986, Application of the induced polarization method to offshore placer resource exploration: Offshore technology conference, 18th Houston, Preprint OTC 5199, p. 395–398.

Wynn, J. C., Grosz, A. E., and Foscz, V. M., 1985, Induced polarization and magnetic response of titanium-bearing placer deposits in the southeastern United States: U.S. Geological Survey Open-File Report 85–756, 25 p.

Yudin, B. A., and Zak, S. I., 1971, Titanium deposits of northwestern USSR (eastern part of Baltic Shield): International Geology Review, v. 13, p. 864–872.

Zen, E-an, 1960, Metamorphism of lower Paleozoic rocks in the vicinity of the Taconic range in west-central Vermont: American Mineralogist, v. 45, p. 129–175.

MANUSCRIPT ACCEPTED BY THE SOCIETY JULY 6, 1990

Typeset by WESType Publishing Services, Inc., Boulder, Colorado
Printed in U.S.A. by JL Printing, Loveland, Colorado